FOOD ENGINEERING

Fundamentals

FOOD ENGINEERING

Fundamentals

First edition (2017)

Arjun Ghimire

To

My beloved family

Preface

Food Engineering is a major contributor to food technology, and provides important and useful tools for the food technologist to apply in designing, developing and controlling food processes. Food engineering principles are the basis for food processing, but only some of them are important and commonly encountered in the food industry. This book aims to select these important principles and show how they can be quantitatively applied in the food industry. It explains, develops and illustrates them at a level of understanding which covers most of the needs of the food technologist in industry and of the student working to become one. This book is also intended to be a step-by-step workbook that will help the students to practice solving food engineering problems.

This book on **"Food Engineering Fundamentals"** can be used to serve as a text or as a reference book for students, professionals, and others engaged in agricultural science and food engineering, food science, and food technology. We are indebted to several authors, whose published materials we took the liberty to use. As a token of thanks, we have appended their work in the bibliography.

The book is intended to introduce technological ideas and concepts on unit operations, and to illustrate their use. Data, including properties and charts, are provided, but for definitive design details may need to be independently checked to ensure requisite precision. Every effort has been made to provide clear explanations and to avoid errors, but errors may occur though.

So feedback from users will be most welcome, and should be directed to the author.

Arjun Ghimire
(Lecturer)
Central Campus of Technology
Tribhuvan University
Dharan, Nepal

Contents

Chapter 1: Material and Energy balances 1
    Introduction 1
    Basic principles of mass and energy balance 1
    Classification of processes 3
    Unit operation 4

Chapter 2: Fluid flow 9
    Volumetric and Mass flow rate 9
    Equation of continuity 9
    Energy in a fluid 12
    Bernoulli's equation 12
    Laminar and turbulent flow 17
    Reynolds' experiment 19
    Flow measuring equipments 21
    Viscosity 31
    Newtonian and non-newtonian fluids 32
    Fanning equation 33
    Friction factors 34
    Moody's diagram 36
    Energy losses in bends and fittings 37

Chapter 3: Heat transfer 39
    Introduction 39
    Mechanism of heat transfer 40
    Fourier's law of heat conduction 41
    Conduction through a flat slab or wall 41
    Conduction through multi-layered slabs in series 44
    Heat conductance in parallel 48
    Conduction through a single hollow cylinder 49
    Conduction through multi-layered cylinders in series 51
    Unsteady state heat transfer 55
    Biot number (Bi) 57
    Lumped heat capacity method (K>>>h) 57
    Fourier number (Fo) 58
    Heat Transfer by convection 58
    Natural convection 58
    Forced convection 60
    Surface heat transfer coefficients 61
    Combined conduction and convection (In a plane slab) 61
    Fouling factors 65
    Heat transfer to boiling liquids 65
    The boiling curve 66

  Heat transfer from condensing vapours 67
  Heat exchangers 68
  Continuous-flow heat exchangers 68
  Log mean temperature difference [LMTD] 70
  NTU method 70
  Shell and tube heat exchangers 70
  Plate heat exchangers 72

Chapter 4: Evaporation 75
  Introduction 75
  Factors affecting evaporation 75
  Boiling-point elevation 76
  Evaporation Equipments 78
  Evaporation for heat-sensitive liquids 81
  Mass and heat balance in single effect evaporator 82
  Classification of evaporators on the basis of operation 87
  Advantages of multiple effect evaporators 89
  Improving the economy of evaporators 90

Chapter 5: Distillation 93
  Introduction 93
  Ideal solution 93
  Raoult's, Dalton's and Henry's law 93
  Volatility and Relative volatility 94
  Boiling point diagram 95
  Methods of Distillation 97
  Equations of operating lines 101
  Effect of feed conditions 104
  Equation of q- line 105
  Location of the feed tray in a tower and number of ideal plates (McCabe-Thiele method) ... 107
  Total reflux ratio 111
  Minimum reflux ratio 112

Chapter 6: Drying 113
  Introduction 113
  Equilibrium moisture content (EMC) 114
  Moisture content representations 115
  Theory of drying 117
  Drying curve 118
  Drying rate curve 119
  Drying equipments 122

Chapter 7: Crystallization 129
  Introduction 129
  Theory of crystallization 129

    Nucleation theory 130
    Rate of crystal growth 131
    L- law of crystal growth 133
    Solubility and saturation 133
    Mass and heat balance 134
    Crystallization processes in the food industries 138
    Classification of crystallizers 139

Chapter 8: Filtration 141
    Introduction 141
    Basic theory of filtration 141
    Constant-rate filtration 143
    Constant-pressure filtration 144
    Filter-cake compressibility 147
    Filtration equipments 147

Chapter 9: Sedimentation 151
    Introduction 151
    The velocity of particles moving in a fluid 151
    Sedimentation 152
    Free and hindered settling 154
    Drag coefficient 155
    Sedimentation equipments 155

Chapter 10: Size reduction 159
    Introduction 159
    Energies used in grinding 159
    Forces used in grinding 163
    Grinding equipments 163
    Screening 168
    Different screen series 169
    Screen openings 170
    Factors affecting the efficiency of screening operation 171

Chapter 11: Mixing 173
    Introduction 173
    Characteristics of mixtures 173
    Measurement of mixing 174
    Mixing of widely different quantities 178
    Rates of mixing 178
    Energy input in mixing 180
    Liquid mixing 180
    Types of mixing 182
    Factors influencing mixing 186

Index..189
Appendices...193
Bibliography..201

List of figures

Fig. 1.1: Mass and energy balance 2
Fig. 1.2: Unit operation 4
Fig. 2.1: Pipe section with constant cross sectional area 9
Fig. 2.3: Energies in a fluid 12
Fig. 2.4: Pipe section with varying cross sectional area and height............................... 13
Fig. 2.5: Reynolds' apparatus 19
Fig. 2.6: Construction of venturimeter 22
Fig. 2.7: Orificemeter.. 27
Fig. 2.8: Construction of pitot tube 28
Fig. 2.9: Construction of rotameter 30
Fig. 2.10: Viscous forces in a fluid 31
Fig. 2.11: Shear stress/shear rate relationships in liquids 33
Fig. 2.13: Moody's diagram 36
Fig. 3.1: Mechanism of heat transfer by (a) conduction (b) convection (c) radiation .. 40
Fig. 3.2: Conduction through a flat slab 41
Fig. 3.3: Conduction through a multi layered slab in series 44
Fig. 3.5: Conduction through a hollow cylinder 49
Fig. 3.6: Conduction through multi-layered cylinders in series..................................... 51
Fig. 3.7: Hot cylinder being cooled in air 56
Fig. 3.8: Combined conduction and convection in a plane slab 61
Fig. 3.9: Boiling curve of water at atmospheric pressure 66
Fig. 3.10: Temperature profiles for heat transfer in a heat exchanger with countercurrent flow . 69
Fig. 3.11: Temperature profiles for heat transfer in a heat exchanger with concurrent flow........ 69
Fig. 3.12: Cross flow heat exchanger... 69
Fig. 3.13: 1 Shell and 1 tube pass heat exchanger 71
Fig. 3.14: 1 Shell and 2 tube pass heat exchanger 71
Fig. 3.15: Principles of flow and heat transfer in a plate heat exchanger 72
Fig. 4.1: Duhring plot for boiling point of sodium chloride solutions........................... 77

Fig. 4.2: Open pan evaporator 78
Fig. 4.3: Horizontal tube evaporator 78
Fig. 4.4: Basket evaporator 79
Fig. 4.5: Climbing film long tube evaporator...80
Fig. 4.6: Forced circulation evaporator 81
Fig. 4.7: Mass and heat balance in a single effect evaporator..................................82
Fig. 4.8: Working of a single effect evaporator 87
Fig. 4.9: Simplified diagram of forward -feed triple-effect evaporator 88
Fig. 4.10: Simplified diagram af backward-feed triple-effect evaporator................89
Fig. 5.1: Boiling point diagram of benzene- toluene mixture at 1 atm 95
Fig. 5.2: Equilibrium diagram for ideal solution at 1 atm (zeotropic) 96
Fig. 5.3: (a) Maximum and minimum boiling azeotrope 97
Fig. 5.4: Differential distillation..98
Fig. 5.5: Flash distillation..99
Fig. 5.6: Working of rectification column..100
Fig. 5.7: Continuous rectification column..101
Fig. 5.8: Enriching section of a rectification column 102
Fig. 5.9: Stripping section of a rectification column 103
Fig. 5.10: Effect of feed conditions 105
Fig. 5.11: Entrance of feed to the rectification column..106
Fig. 5.12: McCabe thiele method for counting number of plates in a rectification column 107
Fig. 5.13: Case of total reflux 111
Fig. 5.14: Case of minimum reflux ratio 112
Fig. 6.1: Equilibrium moisture contents...114
Fig. 6.2: Movement of moisture during drying 117
Fig. 6.3: Typical drying curve 118
Fig. 6.4: Drying rate curve 120
Fig. 6.5: Tray dryer...122
Fig. 6.6: Concurrent tunnel dryer 123
Fig. 6.7: Countercurrent tunnel dryer...124
Fig. 6.8: Centre exhaust tunnel dryer 124
Fig. 6.9: Cross flow tunnel dryer..125
Fig. 6.10: Single stage drum dryer 126
Fig. 6.11: Spray dryer...127
Fig. 6.12: Fluidized bed dryer 128
Fig. 7.1: Mier's theory of crystallization..130

Fig. 7.2: Rate of crystal growth 131
Fig. 7.3: Solubility and saturation curves for sucrose in water 134
Fig. 7.4: Agitated batch crystallizer 139
Fig. 7.5: Vacuum crystallizer 140
Fig. 8.1: Simple laboratory filtration apparatus 141
Fig. 8.2: Basic theory of filtration 141
Fig. 8.3: Plate and frame filter press 148
Fig. 8.4: Rotary filter 149
Fig. 8.5: Centrifugal filter 149
Fig. 8.6: Air filter 150
Fig. 9.1: Motion of particle in a fluid 154
Fig. 9.2: Settler for liquid- liquid dispersion 155
Fig. 9.3: Dust- settling chambers 156
Fig. 9.4: Simple gravity settling classifier 156
Fig. 9.5: Spitzkasten classifier 157
Fig. 9.6: Continuous thickener 157
Fig. 9.7: Centrifugal cream separator 158
Fig. 10.1: Jaw crusher 163
Fig. 10.2: Gyratory crusher 164
Fig. 10.3: Hammer mill 164
Fig. 10.4: Roller mill 165
Fig. 10.5: Single disc attrition mill 166
Fig. 10.6: Double disc attrition mill 166
Fig. 10.7: Burr mill 167
Fig. 10.8: Ball mill 167
Fig. 10.9: Perforated metal screens 170
Fig. 10.10: Wire mesh screens 171
Fig. 11.1: Performance of propeller mixers 181
Fig. 11.2: Mixers 182
Fig. 11.3: Kneader 183
Fig. 11.4: Homogenizing valve 184
Fig. 11.5: Propeller mixer 186

List of tables

Table 2.1: Friction loss factors in fittings 37
Table 3.1: Typical fouling coefficient in W/ m^{2}K 65
Table 6.1: Advantages and limitations of parallel flow, counter-current flow, centre -exhaust and cross-flow drying 125

List of solved numericals

Example 1.1: Concentration of cane sugar 5
Example 1.2: Cooling of pea soup 5
Example 2.1: Calculation of beer velocity 10
Example 2.2: Pressure in a pipeline 15
Example 2.3: Flow regime of milk 20
Example 2.4: Flowrate in venturimeter 24
Example 2.5: Velocity of oil in pitot tube 29
Example 2.6: Pressure drop of olive oil 35
Example 3.1: Heat transfer through a glass window 42
Example 3.2: Heat transfer through a multilayered surface 45
Example 3.3: Heat transfer through a stainless steel pipe 50
Example 3.4: Heat transfer through a multi-layered steel pipe 52
Example 3.5: Overall heat transfer coefficient in slab 63
Example 4.1: Heat transfer area of single effect evaporator 83
Example 5.1: Number of plates in rectification column 108
Example 6.1: Drying of paddy 115
Example 6.2: Drying of wet solid 121
Example 7.1: Solubility of sodium chloride 134
Example 7.2: Crystallization of salt 135
Example 8.1: Constant rate filtration in an air filter 143
Example 8.2: Constant pressure filtration in a filter 145
Example 9.1: Settling velocity of dust particles 153
Example 10.1: Milling of food using Rittinger's equation 161
Example 10.2: Crushing of limestone 161
Example 11.1: Mixing of binary mixtures 175
Example 11.2: Mixing index during dough mixing 177
Example 11.3: Mixing time of binary mixtures 179

List of problems

Mass and energy balances 6
Continuity equation 11
Bernoulli's equation 16
Flow regimes 20
Venturimeter 25
Orificemeter 27
Pitot tube 29
Heat transfer through a single slab 42
Heat transfer through a multilayered slabs 46
Heat transfer through a single layered cylinder 51
Heat transfer through a multilayered cylinders 54
Dimensionless numbers in convection 59
Heat transfer by convection 61
Overall heat transfer coefficient 63
Single effect evaporator 84
Operating sections of rectification 104
q-line equation 107
Calculating number of plates in rectification 110
Moisture content representations 116
Drying times 121
Mass and heat balance in crystallizer 136
Rate of filtration 146
Energy required in size reduction 162
Mixing index 180

List of equations

Mass balance (eq^n 1.1) 2

Energy balance (eq^n 1.2) 3

Volumetric and mass flow rate (eq^n 21. & 2.2) 9

Equation of continuity (eq^n 2.3) 10

Bernoulli's equation (eq^n 2.4) 13

Reynold's number (eq^n 2.5) 18

Flow rates using venturimeter (eq^n 2.11 & 2.12) 23

Practical flowrates using C_D (eq^n 2.13 & 2.14) 24

Flow rate using pitot tube (eq^n 2.16) 28

Fundamental equation of viscosity (eq^n 2.18) 32

Fanning equation (eq^n 2.24) 34

Energy loss in bends and fittings (eq^n 2.26) 37

Energy loss at sudden enlargement (eq^n 2.27) 38

Energy loss at sudden contraction (eq^n 2.28) 38

Fourier's law of heat conduction (eq^n 3.1) 41

Conduction through a flat slab (eq^n 3.2) 42

Conduction through multilayered slabs in series (eq^n 3.7) 45

Conduction through a single hollow cylinder (eq^n 3.11) 50

Conduction through multi-layered cylinders in series (eq^n 3.15) 52

Unsteady state heat transfer (eq^n 3.20) 56

Lumped heat capacity (eq^n 3.23) 57

Fourier number (eq^n 3.24) 58

Dimensionless numbers in convection (eq^n 3.25, 3.26 & 3.27) 59

Surface heat transfer coefficients (eq^n 3.29) 61

Overall heat transfer coefficient (eq^n 3.35) 62

Fouling factor (eq^n 3.37) 65

Heat transfer from condensing vapors (eq^n 3.38 & 3.39) 67

Log mean temperature difference (eq^n 3.40) 70

Number of transfer units (eq^n 3.42) 70

Boiling point elevation (eq^n 4.1) 76

The overall rate of heat transfer for single effect evaporator (eq^n 4.6) 83

Raoult's, Dalton's and Henry's law (eq^n 5.1, 5.2 & 5.3) 94

Volatility (eq^n 5.4) 94

Relative volatilty (eq^n 5.5) 95

Rayleigh equation (eq^n 5.12) 98
Equation for enriching section (eq^n 5.19) 102
Equation for stripping section (eq^n 5.20) 104
q- line equation (eq^n 5.28) 106
Moisture content on wet basis (eq^n 6.1) 115
Moisture content on dry basis (eq^n 6.2) 115
Drying times (eq^n 6..4 & 6.5) 120
Rate of crystal growth (eq^n 7.4) 132
L- law of crystal growth (eq^n 7.5) 133
Fundamental equation for filtration (eq^n 8.5) 142
Constant rate filtration (eq^n 8.6) 143
Constant pressure filtration (eq^n 8.8) 145
Specific resistance of filter cake (eq^n 8.10) 147
Fundamental equation for movement of particles in fluids (eq^n 9.5) 152
Stoke's law (eq^n 9.6) 152
Energy required in grinding (eq^n 10.1) 159
Rittinger's law (eq^n 10.3) 159
Kick's law (eq^n 10.5) 160
Bond's law (eq^n 10.6) 160
Variance (eq^n 11.2) 176
Mixing index (eq^n 11.5) 177
Rate of mixing (eq^n 11.7) 178
Power number (eq^n 11.8) 180

Chapter 1: Material and Energy balances

Introduction

Material quantities, as they pass through food processing operations, can be described by material balances. Such balances are statements on the conservation of mass. Similarly, energy quantities can be described by energy balances, which are statements on the conservation of energy. If there is no accumulation, what goes into a process must come out. This is true for batch operation. It is equally true for continuous operation over any chosen time interval.

Material and energy balances are very important in the food industry. Material balances are fundamental to the control of processing, particularly in the control of yields of the products. The first material balances are determined in the exploratory stages of a new process, improved during pilot plant experiments when the process is being planned and tested, checked out when the plant is commissioned and then refined and maintained as a control instrument as production continues. When any changes occur in the process, the material balances need to be determined again.

The increasing cost of energy has caused the food industry to examine means of reducing energy consumption in processing. Energy balances are used in the examination of the various stages of a process, over the whole process and even extending over the total food production system from the farm to the consumer's plate.

Material and energy balances can be simple, at times they can be very complicated, but the basic approach is general. Experience in working with the simpler systems such as individual unit operations will develop the facility to extend the methods to the more complicated situations, which do arise. The increasing availability of computers has meant that very complex mass and energy balances can be set up and manipulated quite readily and therefore used in everyday process management to maximize product yields and minimize costs.

Basic principles of mass and energy balance

If the unit operation, whatever its nature is seen as a whole it may be represented diagrammatically as a box, as shown in the Fig. 1.1. The mass and energy going into the box must balance with the mass and energy coming out.

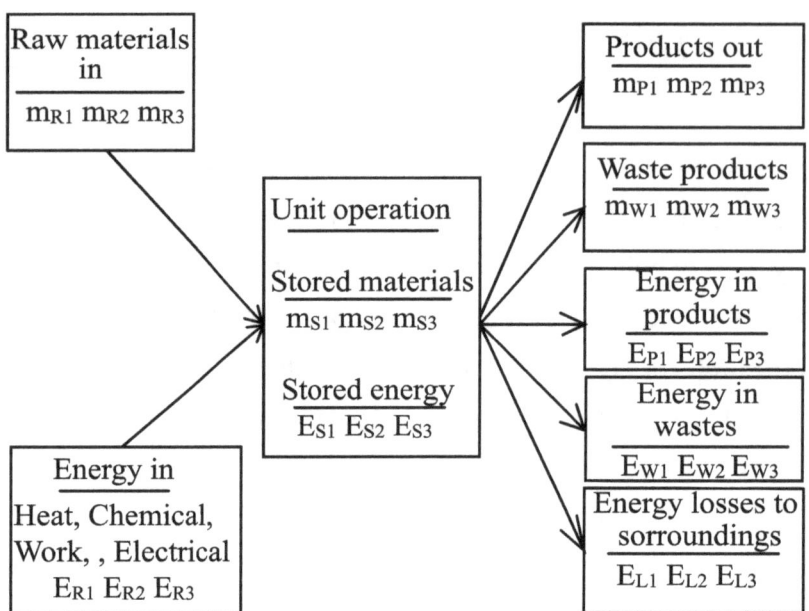

Fig. 1.1: Mass and energy balance

The law of conservation of mass leads to what is called a mass or a material balance.

Mass In = Mass Out + Mass Stored………. (1.1)

Raw Materials = Products + Wastes + Stored Materials.

$\sum m_R = \sum m_P + \sum m_W + \sum m_S$

   (where $\sum$ (sigma) denotes the sum of all terms).

$\sum m_R = m_{R1} + m_{R2} + m_{R3} + \ldots$ = Total Raw Materials.

$\sum m_P = m_{P1} + m_{P2} + m_{P3} + \ldots$ = Total Products.

$\sum m_W = m_{W1} + m_{W2} + m_{W3} + \ldots$ = Total Waste Products.

$\sum m_S = m_{S1} + m_{S2} + m_{S3} + \ldots$ = Total Stored Materials.

If there are no chemical changes occurring in the plant, the law of conservation of mass will apply also to each component, so that for component A:

   m_A in entering materials = m_A in the exit materials + m_A stored in plant.

For example, in a plant that is producing sugar, if the total quantity of sugar going into the plant in sugar cane or sugar beet is not equaled by the total of the purified sugar and the sugar in the waste liquors, then there is something wrong. Sugar is either being burned (chemically changed) or accumulating in the plant or else it is going unnoticed down the drain somewhere. In this case:

$(m_A) = (m_{AP} + m_{AW} + M_{AS} + m_{AU})$

where m_{AU} is the unknown loss and needs to be identified. So the material balance is now:

Raw Materials = Products + Waste Products + Stored Products + Losses where Losses are the unidentified materials.

Just as mass is conserved, so is energy conserved in food processing operations. The energy coming into a unit operation can be balanced with the energy coming out and the energy stored.

Energy In = Energy Out + Energy Stored………. (1.2)

$\sum E_R = \sum E_P + \sum E_W + \sum E_L + \sum E_S$

where,

$\sum E_R = E_{R1} + E_{R2} + E_{R3} + \ldots\ldots$ = Total Energy Entering
$\sum E_P = E_{P1} + E_{P2} + E_{P3} + \ldots\ldots$ = Total Energy Leaving with Products
$\sum E_W = E_{W1} + E_{W2} + E_{W3} + \ldots\ldots$ = Total Energy Leaving with Waste Materials
$\sum E_L = E_{L1} + E_{L2} + E_{L3} + \ldots\ldots$ = Total Energy Lost to Surroundings
$\sum E_S = E_{S1} + E_{S2} + E_{S3} + \ldots\ldots$ = Total Energy Stored

Energy balances are often complicated because forms of energy can be interconverted, for example mechanical energy to heat energy, but overall the quantities must balance.

Classification of processes

Based on how the process varies with time

Steady-state process is one where none of the process variables change with time. Every time we take a snapshot, all the process variables have the same values as in the first snapshot.

Unsteady-state (Transient) process is one where the process variables change with time. Every time we take a snapshot, many of the variables have different values than in the first snapshot.

One class of unsteady-state processes is oscillatory, where the process variables change with time in a regular way. All other unsteady processes may be called Transient meaning that the process variables continuously evolve over time.

Based on how the process was built to operate

A Continuous process is one that has the feed streams and product streams moving into and out of the process all the time. Examples are an oil refinery, a power grid, pool filter and a distillation process.

A Batch process is a process one, where the feed streams are fed to the process to get it started. The feed material is then processed through various process steps and the finished products are taken out at specific times.

Steps:
- Feed is charged into vessel
- Process is started
- No mass is added or removed from vessel (temperature and pressure are usually monitored and controlled)
- At some conditions or fixed time, products are removed

In a batch process no material is exchanged with the surroundings during the process. Examples: baking cookies, fermentations, small-scale chemicals (pharmaceuticals).

A Semi-batch process (also called semi-continuous) is a process that has some characteristics continuous and batch processes. Examples are washing machine, fermentation with purge etc.

Unit operation

Using a material balance and an energy balance, a food engineering process can be viewed overall or as a series of units. Each unit is a unit operation. The unit operation can be represented by a box as shown in Fig. 1.2.

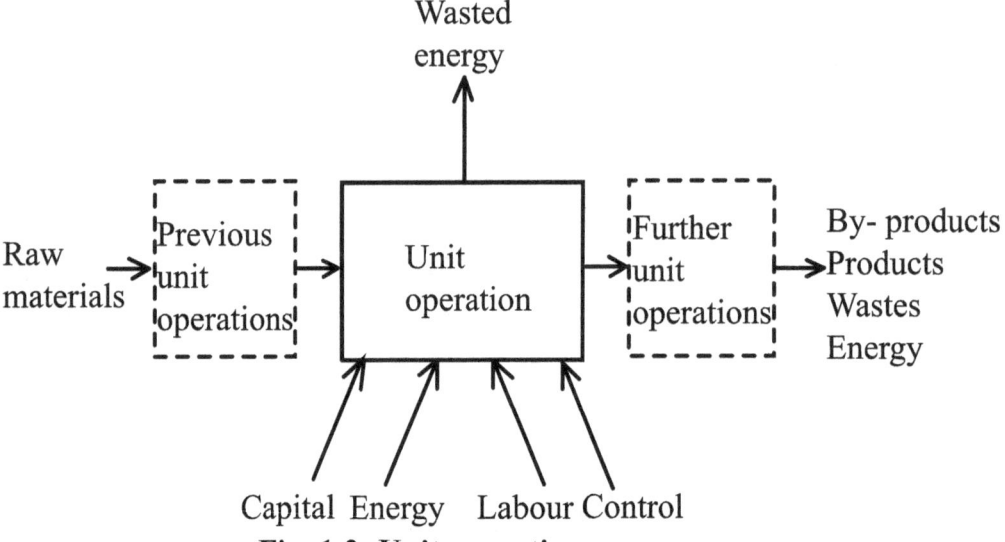

Fig. 1.2: Unit operation

Into the box go the raw materials and energy, out of the box come the desired products, by-products, wastes and energy. The equipment within the box will enable the required changes to be made with as little waste of materials and energy as possible. In other words, the desired products are required to be maximized and the undesired by-products and wastes minimized. Control over the process is exercised by regulating the flow of energy, or of materials, or of both.

Example 1.1: Concentration of cane sugar

1. An evaporator is used to concentrate cane sugar solution. A feed of 10000kg/hr of a solution 38% wt. sugar is evaporated producing a 74% wt. sugar solution. Calculate the weight of sugar produced and the amount of water evaporated.

Sol^n:

i. Draw the process diagram:

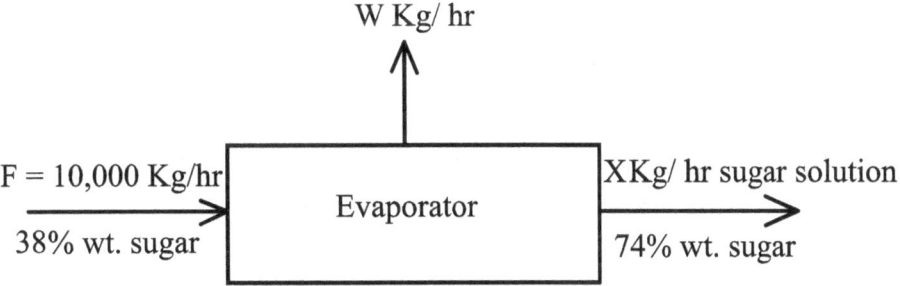

ii. Make overall Material balance on the system:

i.e. $F = W + X$

Or, $10{,}000 = W + X$ .... (1.3)

iii. Make a component material balance (Sugar):

i.e. Sugar content in the feed = Sugar content in the evaporated water + Sugar content in the exit

Or, $\left(\dfrac{38}{100}\right) \times 10{,}000 = 0 \times W + \left(\dfrac{74}{100}\right) \times X$

As, solid content in evaporated water is zero.

Or, $0.74\, X = 3800$

Or, $X = 5135.135$ Kg/ hr.

Hence, from equation (1.3)

Amount of water evaporated, $W = 4864.86$ Kg/ hr &

The weight of sugar produced, $X = 5135.135$ Kg/ hr.

Example 1.2: Cooling of pea soup

An autoclave contains 1000 cans of pea soup. It is heated to an overall temperature of 100°C. If the cans are to be cooled to 40°C before leaving the autoclave, how much cooling water is required, if it enters at 15°C and leaves at 35°C? Note the additional information below:

The specific heats of the pea soup and the can metal are respectively 4.1 kJ/ kg°C and 0.50 kJ/

kg°C. The weight of each can is 60g and it contains 0.45 kg of pea soup. Assume that the heat content of the autoclave walls above 40°C is 16000 kJ and that there is no heat loss through the walls.

Sol^n:

Let w = the weight of cooling water required; and the datum temperature be 40°C, the temperature of the cans leaving the autoclave.

<u>Heat entering</u>

1) Heat in cans = weight of cans × specific heat × temperature above datum
$$= 1000 × 0.06 × 0.50 × (100-40) \text{ kJ} = 1800 \text{ kJ}$$
2) Heat in can contents = weight pea soup × specific heat × temperature above datum
$$= 1000 × 0.45 × 4.1 × (100 - 40) = 110700 \text{ kJ}$$
3) Heat content in the autoclave = 16000 kJ
4) Heat in water = weight of water × specific heat × temperature above datum
$$= w × 4.186 × (15-40) = -104.6 \text{ w kJ}.$$
5) Total Heat Entering = 1800 + 110700 + 16000 – 104.6 w = 128500-104.6 w.......... **(1.4)**

<u>Heat leaving</u>

1) Heat in cans = 1000 × 0.06 × 0.50 × (40-40) (cans leave at datum temperature) = 0
2) Heat in can contents = 1000 × 0.45 × 4.1 × (40-40) = 0
3) Heat leaving the autoclave = 0
4) Heat in water = w × 4.186 × (35-40) = -20.9 w
5) Total heat leaving = -20.9 w.......... **(1.5)**

<u>Heat - Energy balance of cooling process; 40°C as datum line, using equation 1.4 & 1.5:</u>

Total heat entering = Total heat leaving

Or, 128500-104.6 w = -20.9 w

$$\therefore w = 1535.24 \text{ kg}$$

Hence, the amount of cooling water required = 1535.24 kg.

Exercises on mass and energy balances

1. How much dry sugar must be added in 100 kg of aqueous sugar solution in order to increase its concentration from 20% to 50%? [**Ans:** 60 kg]

2. In a process producing jam, crushed fruit containing 14% wt. of soluble solids are mixed in a mixer with sugar (1.22 kg sugar/ kg crushed fruit) and pectin (0.0025 kg pectin/ kg fruit). The

resultant mixture is then evaporated in a kettle to produce a jam containing 67% wt. soluble solids. For a feed of 1000kg crushed fruits, calculate the kg. mixture from the mixture, kg water evaporated and kg jam produced. [**Ans:** 2222.5, 189 and 2033.42 kg]

3. Fresh orange juice with 12% soluble solids content is concentrated to 60% in a multiple effect evaporator. To improve the quality of the final product, the concentrated juice is mixed with an amount of fresh juice (cut back) so that the concentration of the mixture is 42%. Calculate how much water per hour must be evaporated in the evaporator, how much fresh juice per hour must be added back and how much final product will be produced if the inlet feed flow rate is 10,000 kg/hr fresh juice. Assume steady state. [**Ans:** 8000, 1200 and 3200 kg/hr]

4. Tapioca is used in many countries for making bread and similar products. The flour is made by drying cassava granules of cassava roots containing 66% wt. of mixture to 5% moisture and then grinding to produce flour. How many kg of granules must be dried and how much water should be removed to produce 5000 kg/hr? [**Ans:** 13970.6 and 8970.58 kg/hr]

5. Tomato juice flowing through a pipe at the rate of 100kg/hr is rated by adding saturated salt solution (26% salt) to the pipeline at a constant rate. At what rate would the saturated salt solution be added to produce 2% salt to the product? [**Ans:** 8.33kg/ min]

6. Fishes are processed into fish meal and used as a supplementary protein food. In the processing, the oil is first extracted to produce wet fish cake containing 80% wt. water and 20% wt. bone dry cake. The wet cake feed is dried in a rotary drum dyer to give a dry fish cake product containing 40% wt. water. Finally the product is finely grinded and packed. Calculate the kg/hr of wet cake needed to produce 1000 kg/hr of dry fish cake product. [**Ans:** 3000kg/hr]

7. An evaporator has a evaporation capacity of 500 kg/hr. Calculate the rate of production of juice concentrate containing 45% total solids from raw juice containing 12% total solids. [**Ans:**181.8 kg/hr]

8. The extract containing 4% solids is feed to an evaporator at the rate of 1000kg/hr and discharged as tea concentrate having 42% solids. Calculate flow rate of outlet stream (concentrate and water). [**Ans:** 95.23 and 904.76 kg/hr]

9. How much weight reduction could result when a material is dried from 80% moisture to 50% moisture? [**Ans:** 60%]

10. How much water should be evaporated from 5% salt solution in order to form 20% salt solution? [**Ans:** 75%]

11. In the concentration of orange juice, a fresh extract of strained juice containing 7.08% of solids is fed to a vacuum evaporator and is concentrate to 58% solids. For 1000kg/hr entering, calculate the amount of outlet stream i.e. concentrated juice and water. [**Ans:** 122.06 and 877.94 kg/hr]

12. Determine the amount of juice concentrate containing 55% solids and single strength juice containing 15% solids that must be mixed to produce 100kg of a concentrate having 45% solids. [**Ans:** 75 and 25 kg]

13. Draw a diagram and set up equations representing the mass balance and component balance for a system involving mixing of pork (15% protein, 20% fat and 63% water) and fat (15% water, 50% fat, 30% protein) to make 100kg of a mixture containing 25% fat. [**Ans:** 16.66 kg]

14. KNO_3 solution of 20% is pumped in an evaporator at the rate of 1000kg/hr. The evaporator is at 149°C and a part of water is evaporated and then the solution is concentrated to 50% and then it is pumped to a crystallizer at 38°C where it crystallizes with 4% water and leaves the system a solution of 37.5% is refluxed. Find out the amount of product, water evaporated, intermediate product after evaporator and the amount recirculated. [**Ans:** 208.33, 791.67, 974.98 and 762.65 kg/hr]

15. 1000kg/hr of a fruit juice with 10% solids is freeze concentrated to 40% solids. The dilute juice is fed to the freezer where the ice crystals are formed and then slush (mixture of ice and juice water) is separated in a centrifugal separator into ice crystals and concentrated juice. The amount of 5000kg/hr of liquid is recycled from the separator to the freezer. Calculate the amount of ice that is removed in a separator and the amount of concentrated juice produced. Assume steady state. [**Ans:** 750 and 250 kg/hr]

16. Strawberries contain about 15 wt% solids and 85 wt% water. To make strawberry jam, crushed strawberries and sugar are mixed in a 45:55 mass ratio and the mixture is heated to evaporate water until the residue contains one –third water by mass. Draw and label a flowchart of this process and calculate how many pounds of strawberries are needed to make a pound of jam. [**Ans:** 0.486 pounds]

Chapter 2: Fluid flow

Volumetric and Mass flow rate

Consider a pipeline with constant cross-sectional area 'A'. The fluid is flowing inside with constant velocity 'v'. Consider the small study segment on the pipeline YY'-XX' of length 'L' as shown in Fig. 2.1 below.

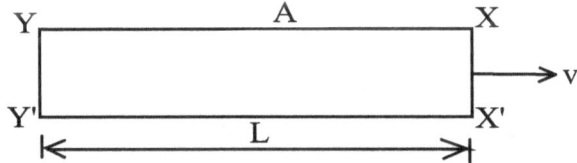

Fig. 2.1: Pipe section with constant cross sectional area

Volume of fluid in element, V= AL

Velocity of fluid, $v = \dfrac{L}{t}$

Since, volumetric flow rate (Q*) is the volume of fluid flowing per unit time,

$$Q^* = \dfrac{V}{t} = \dfrac{A.L}{t} = Av \ (m^3/s) \ \dots\dots\dots \ (2.1)$$

Similarly, Mass of fluid in element, $m = \rho V = \rho.AL$

Since, mass flow rate (m*) is the mass of fluid flowing per unit time,

$$m^* = \dfrac{m}{t} = \dfrac{\rho AL}{t} = \rho A v$$

$$m^* = \rho \ Q^* \ (kg/s) \ \dots\dots\dots \ (2.2)$$

Equation of continuity

It is based on the law of conservation of mass to obtain a mass balance. Once the system is working steadily, and if there is no accumulation of fluid in any part the system, the quantity of fluid that goes in at section 1 must come out at section 2.

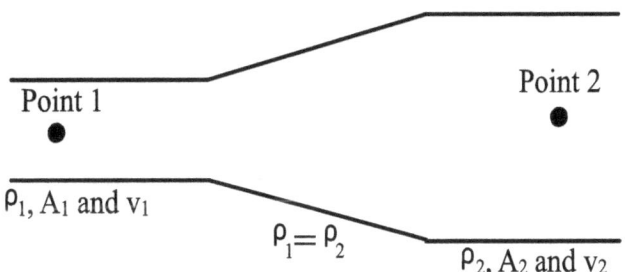

Fig. 2.2: Pipe section with varying cross sectional area

If the area of the pipe at section 1 is A_1, the velocity at this section v_1 and the fluid density ρ_1; and if the corresponding values at section 2 are A_2, v_2, ρ_2 as shown in Fig. 2.2.

Based on law of conservation of mass:

i.e. $m_1^* = m_2^*$

Or, $\rho_1 A_1 v_1 = \rho_2 A_2 v_2$

Or, $A_1 v_1 = A_2 v_2$, as the fluid is incompressible i.e, $\rho_1 = \rho_2$

Hence, $\boxed{Q_1^* = Q_2^*}$ .... (2.3)

Example 2.1: Calculation of beer velocity

1. The Volumetric flow rate of beer flowing in a pipe is 1.8 L/s. The inside diameter of the pipe is 3cm. The density of beer is 1100 kg/m³. Calculate the average velocity of beer and its mass flow rate in kg/s. If another pipe with a diameter of 1.5cm is used, what will be the velocity for the same volumetric flow rate?

Sol^n:

Given,

Volumetric flowrate of beer (Q^*) = 1.8 L/s = 0.0018 m³/s

Inside diameter of the tube (d) = 3 cm = 0.03 m

Density of beer (ρ) = 1100 Kg/m³

Therefore, Cross-sectional area of pipe (A) = $\dfrac{\pi d^2}{4} = \dfrac{3.14 \times (0.03)^2}{4} = 0.0007065$ m²

Hence, from equation 2.1;

i. velocity of beer (v) = $\dfrac{Q^*}{A} = \dfrac{0.0018}{0.0007065} = 2.55$ m/s.

Using equation 2.2;

ii. Mass flow rate of beer (m*) = $\rho \times Q^* = 1100 \times 0.0018 = 1.98$ Kg/s.

iii. When a new diameter tube of 1.5 cm is used, the cross sectional area of tube becomes

$(A_n) = \dfrac{3.14 \times (0.015)^2}{4} = 0.000176625$ m²

Therefore, the velocity now changes to $(v_n) = \dfrac{Q^*}{A_n} = \dfrac{0.0018}{0.000176625} = 10.19$ m/s.

Hence, the new velocity is found to be 10.19 m/s.

Exercises on continuity equation

1. Saturated steam of specific volume 0.3928m³/kg is flowing in a steel pipe of internal diameter 4.925cm. If the average velocity of the steam is 10m/s, calculate the mass flow rate of the steam. [**Ans:** 0.0485 kg/s]

2. Water is flowing at a rate of 2m/s in a pipe of cross-sectional area 0.02m². If the cross-section is reduced to half, then find the rate of flow. [**Ans:** 4m/s]

3. On a circular conduit, the diameter changes from 2m to 3m. The velocity in the entrance profile was measured as 3 ms^{-1}. Calculate the discharge and mean velocity at the outlet profile. [**Ans:** 9425 m³s^{-1} and 1333ms^{-1}]

4. If the water that exits a pipe fills a pool that is 3 meters deep, 20 meters long, and 5 meters wide in 3 days, what is the flow rate? [**Ans:** 1.2×10^{-3}m³s^{-1}]

5. A nozzle of inner radius 2mm is connected to a hose of inner radius 4mm. The nozzle shoots out water moving at 20m/s. i. At what speed is the water in the hose moving? ii. What is the volumetric flow rate? iii. What is the mass flow rate? [**Ans:** 5m/s, 2.51×10^{-4} m³/s and 0.251kg/s]

6. A sucrose solution is flowing in a pipe with 0.0475m inside diameter and 10m length at the rate of 3m³/h. Find the mean velocity of the liquid. [**Ans:** 0.471m/s]

7. A large artery in a dog has an inner radius of 4.00×10^{-3} m. Blood flows through the artery at the rate of 1.00×10-6 m³s^{-1}. The blood has a viscosity of 2.084×10^{-3} Pa.s and a density of 1.06×10^{3} kg.m^{-3}. Calculate the average blood velocity in the artery. [**Ans:** 1.99×10^{-2} ms^{-1}]

8. The radius of the aorta is about 1cm and the blood flowing through it has a speed of about 30 cm/s. Calculate the average speed of the blood in capillaries given that, although each capillary has a diameter of about 8×10^{-4} cm, there are literally billions of them so that their total cross-section area is about 2000 cm². [**Ans:** 5×10^{-4} m/s]

9. A 3m diameter stainless steel tank contains wine. In the tank, the wine is filled to 5m depth. A discharge port, 10 cm diameter is opened to drain the wine. Calculate the discharge velocity of wine assuming the flow is steady and frictionless, and the time required in emptying it. [**Ans:** 0.078m³/s and 7.5 mins]

Energy in a fluid
There are 4 types of energy possessed by a fluid as mentioned in Fig. 2.3.

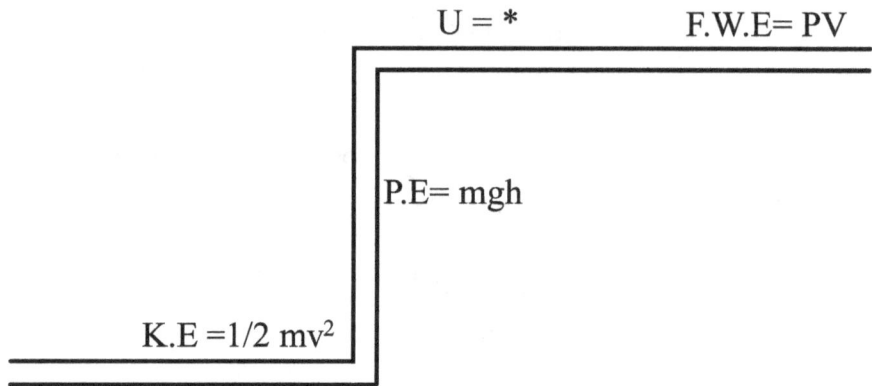

Fig. 2.3: Energies in a fluid

Kinetic energy (K.E) is the energy due to velocity as a whole.

$$K.E = \frac{1}{2}mv^2 \quad \text{[m is mass and v is velocity of fluid]}$$

Potential energy (P.E) is the energy due to height of the fluid.

P.E= mgh   [g is acceleration due to gravity and h is height of fluid]

Internal energy (U) is the energy due to temperature of the fluid.

U = *. It is not assumed as the temperature change is considered zero.

Flow work energy (F.W.E) is the energy required to displace the fluid.

F.W.E= PV   [P is pressure and V is volume of the fluid]

Bernoulli's equation

It is based on the law of conservation of energy. The Bernoulli Equation can be considered to be a statement of the conservation of energy principle appropriate for flowing fluids. The qualitative behavior that is usually labeled with the term "Bernoulli effect" is the lowering of fluid pressure in regions where the flow velocity is increased.

In Fig. 2.4, the total energy of fluid entering at section 1 must equal to the total energy of fluid leaving at section 2.

<u>Assumptions</u>
1. No heat transfer (Adiabatic)
2. No work done by pumps or turbines.
3. Flow is frictionless (No temperature change)
4. Flow is incompressible (Constant density)

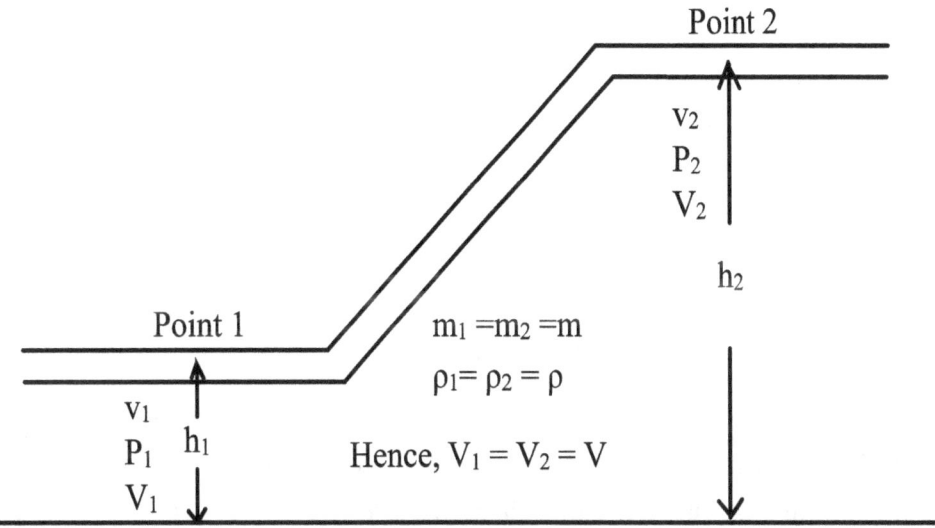

Fig. 2.4: Pipe section with varying cross sectional area and height

<u>At steady state</u>

Total energy of fluid at point 1 = Total energy at point 2

i.e. $K.E_1 + P.E_1 + F.W.E_1 = K.E_2 + P.E_2 + F.W.E_2$

Or, $\frac{1}{2}mv_1^2 + mgh_1 + P_1V_1 = \frac{1}{2}mv_2^2 + mgh_2 + P_2V_2$

For incompressible fluid; $V_1 = V_2 = V$

Dividing both sides by V:

$$P_1 + \frac{1}{2V}mv_1^2 + \frac{mgh_1}{V} = P_2 + \frac{1}{2V}mv_2^2 + \frac{mgh_2}{V} = \text{Constant}$$

Since density, $\rho = \frac{m}{V}$

Or, $P_1 + \frac{1}{2}\rho v_1^2 + \rho gh_1 = P_2 + \frac{1}{2}\rho v_2^2 + \rho gh_2 = \text{Constant}\ldots\ldots\ldots$ **(2.4)**

Hence, $\boxed{\frac{1}{2}\rho v^2 + \rho gh + P = \text{Constant} = \text{Total Pressure}}$

We note that the pressure of the system is constant in this form of the Bernoulli Equation. If the static pressure of the system (the far right term) increases, and if the pressure due to elevation (the middle term) is constant, then we know that the dynamic pressure (the left term) must have decreased. In other words, if the speed of a fluid decreases and it is not due to an elevation difference, we know it must be due to an increase in the static pressure that is resisting the flow.

Applications of Bernoulli's theorem

1. When we blow air over a strip of paper as shown in the below figure, we find that the paper moves up. This is because, on blowing air, the velocity of air increases, creating low pressure above the paper and high pressure below the paper. This difference in pressure, lifts the paper.

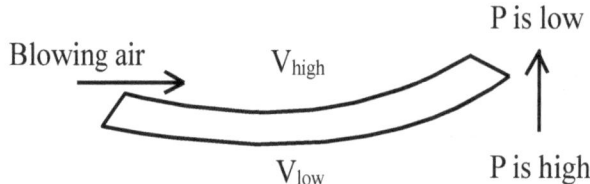

2. The working of spray-gun is based on Bernoulli's theorem. When the rubber bulb is squeezed, air is blown into the tube A, due to which, low pressure and high velocity is created. Since this pressure is less than the atmospheric pressure, the liquid is pushed up. This rising liquid is sprayed out of the nozzle 'N', due to the blowing air.

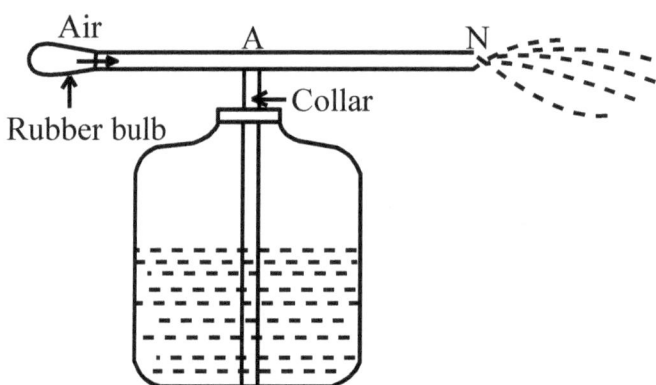

3. The wings of aero plane are designed such that the upper surface has a greater curvature than its lower surface. As the aero plane moves, the air blows in the form of streamlines. The air travels a longer distance over the upper surface when compared to distance to its lower surface, in a given time. This creates a difference in the velocity of the air above and below, as shown. The pressure at top of the wing is low and below the wing is high. This pressure difference gives it an additional thrust.

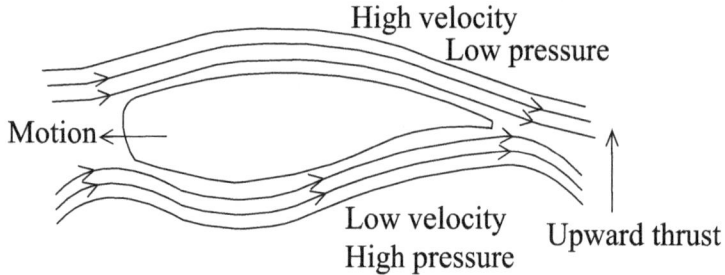

Example 2.2: Pressure in a pipeline

1. Water flows at the rate of 0.4 m³ min⁻¹ in a 7.5 cm diameter pipe at a pressure of 70 KPa. If the pipe reduces to 5 cm diameter calculate the new pressure in the pipe. Density of water is 1000 kg m⁻³.

Sol^n:

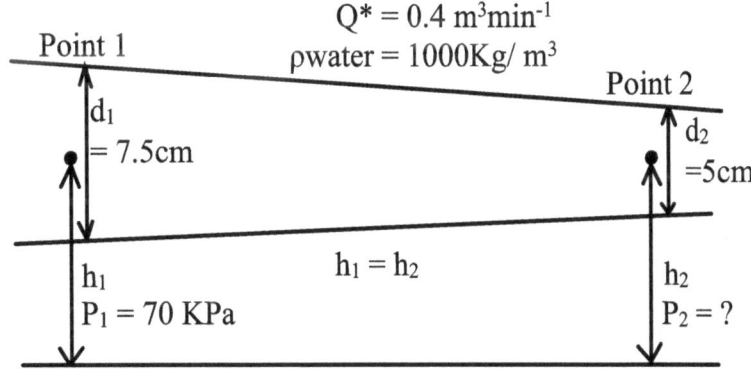

Given,
Volumetric flow rate of water (Q^*) = 0.4 m³/ min = 0.00666 m³/s.

At point 1:
Diameter of the pipe (d_1) = 7.5 cm = 0.075 m
Pressure of water (P_1) = 70 KPa = 70,000 Pa

Cross-sectional area of the pipe (A_1) = $\dfrac{\pi d_1^2}{4}$ = 0.004415 m².

Therefore, velocity of water at point 1 (v_1) = $\dfrac{Q^*}{A_1}$ = 1.508 m/s

At point 2:
Diameter of the pipe (d_2) = 5 cm = 0.05 m
Pressure of water (P_2) =?

Cross-sectional area of the pipe (A_2) = $\dfrac{\pi d_2^2}{4}$ = 0.00196 m².

Therefore, velocity of water at point 2 (v_2) = $\dfrac{Q^*}{A_2}$ = 3.3979 m/s

Applying Bernoulli's' equation, from equation 2.4;

Total energy of fluid at point 1 = Total energy at point 2

Or, $P_1 + \frac{1}{2}\rho v_1^2 + \rho g h_1 = P_2 + \frac{1}{2}\rho v_2^2 + \rho g h_2$

Or, $P_1 + \frac{1}{2}\rho v_1^2 = P_2 + \frac{1}{2}\rho v_2^2$ [Since $h_1 = h_2$]

Or, $71137.032 = P_2 + 5746.05$

Or, $P_2 = 65,390.982$ Pa $= 65.39$ KPa.

Exercises on Bernoulli's equation

1. A hose lying on the ground has water coming out of it at a speed of 5.4 meters per second. The nozzle of the hose is lifted to a height of 1.3 meters above the ground. At what speed does the water now come out of the hose? (**Ans:** 1.9 m/s)

2. Water having density of 998 kg/m³ is flowing at the rate of 1.676m/s in a 3.068 inch diameter horizontal pipe at a pressure P_1 of 68.9 KPa abs. It then passes to a pipe having an inside diameter of 2.069 inch. Calculate the new pressure P_2 in the 2.069 inch pipe. Assume no friction losses. [**Ans:** 63.5 KPa].

3. The diameter of a pipe changes from 200mm at a section 5m above datum to 50 mm at a section 3m above the datum. The pressure of water at the first section is 500KPa. If the velocity of flow at the first section is 1m/s, determine the intensity of pressure at the second section. [**Ans:** 392.4KPa]

4. Water flows through an S-shaped pipe. At one end, the water in the pipe has a pressure of 150,000 Pascal (Pa), a speed of 5.0 m/s, and a height of 0.0 m. At the other end, the speed of the water is 10 m/s, and the height is now 2.0 m. Since the density of water is 1000 kg/m³, calculate the pressure at the exit end of the pipe. [**Ans:** 92,900 Pa]

5. Water is flowing in a fire hose with a velocity of 1.0 m/s and a pressure of 200000 Pa. At the nozzle the pressure decreases to atmospheric pressure (101300 Pa), there is no change in height. Use the Bernoulli equation to calculate the velocity of the water exiting the nozzle. [**Ans:** 14m/s]

6. Through a refinery, fuel ethanol is flowing in a pipe at a velocity of 1 m/s and a pressure of 101300 Pa. The refinery needs the ethanol to be at a pressure of 2 atm (202600 Pa) on a lower level. How far must the pipe drop in height in order to achieve this pressure? Assume the velocity

does not change. (Hint: The density of ethanol is 789 kg/m³ and gravity g is 9.8 m/s²) [**Ans:** -13.1m]

7. A pipe of diameter 400 mm carries water at a velocity of 25m/s. The pressure at the points A and B are given as 29.43 N/cm² and 22.563 N/cm² respectively while the datum head at A and B are 28m and 30m. Find the loss of heat between A and B. [**Ans:** 49070 J]

8. A plane is flying at an altitude of 1000 m above the ground. The density of the air is 1.123 kg/m³, the velocity of the air on the top of the wing is 60 m/s, the velocity on the bottom of the wing is 30 m/s, and the pressure on top of the wing is 88,600 Pa. Find the pressure on the bottom of the wing. [**Ans:** 90112 Pa]

9. The following are the data given of a change in diameter, effected in laying a water supply pipe line. The change in diameter is gradual from 200 mm at A to 500 mm at B. Pressure at A and B are 78.5 kN/m² and 58.9 kN/m² respectively with the end B being 3 m higher than A. If the glow in the pipe line is 200 L/s, find: a) direction of flow and b) the head lost in friction between A and B. [**Ans:** A to B, 0.9345 m]

10. A pipe 400 mm diameter carries water at a velocity of 2.5 m/s. The pressure head at pints A and B are given as 30 m and 23 m respectively, while the datum head at A and B are 28 m and 30 m respectively. Find the loss of head between A and B. [**Ans:** 5 m]

Laminar and turbulent flow

When a liquid flowing in a pipe is observed carefully, it will be seen that the pattern of flow becomes more disturbed as the velocity of flow increases. Perhaps this phenomenon is more commonly seen in a river or stream. When the flow is slow the pattern is smooth, but when the flow is more rapid, eddies develop and swirl in all directions and at all angles to the general line of flow.

At the low velocities, flow is calm. In a series of experiments, Reynolds showed this by injecting a thin stream of dye into the fluid and finding that it ran in a smooth stream in the direction of the flow. As the velocity of flow increased, he found that the smooth line of dye was broken up until finally, at high velocities, the dye was rapidly mixed into the disturbed flow of the surrounding fluid.

From analysis, which was based on these observations, Reynolds concluded that this instability of flow could be predicted in terms of the relative magnitudes of the velocity and the viscous forces that act on the fluid. In fact the instability which leads to disturbed, or what is called "turbulent" flow, is governed by the ratio of the kinetic and the viscous forces in the fluid stream. The kinetic (inertial) forces tend to maintain the flow in its general direction but as they increase so does instability, whereas the viscous forces tend to retard this motion and to preserve order and reduce eddies.

The inertial force is proportional to the velocity pressure of the fluid ρv^2 and the viscous drag is proportional to $\mu v/D$ where D is the diameter of the pipe. The ratio of these forces is:

$$\frac{\rho v^2 D}{\mu v} = \frac{D v \rho}{\mu} = R_e \ldots\ldots\ldots (2.5)$$

This ratio is very important in the study of fluid flow. As it is a ratio, it is dimensionless and so it is numerically independent of the units of measurement so long as these are consistent. It is called the Reynolds number and is denoted by the symbol (R_e).

From a host of experimental measurements on fluid flow in pipes, it has been found that the flow remains calm or "streamline" for values of the Reynolds number up to about 2100. For values above 4000 the flow has been found to be turbulent. Between above 2100 and about 4000 the flow pattern is unstable; any slight disturbance tends to upset the pattern but if there is no disturbance, streamline flow can be maintained in this region.

<u>To summarise for flow in pipes:</u>
For (R_e) < 2100 streamline flow,
For 2100 < (R_e) < 4000 transition,
For (R_e) > 4000 turbulent flow.

Heat transfer occurs at the channel wall. Laminar flow develops an insulating blanket around the channel wall and restricts heat transfer. Conversely, turbulent flow, due to the agitation factor, develops no insulating blanket and heat is transferred very rapidly. Turbulent flow occurs when the velocity in a given water channel is high. Although too much velocity can cause erosion. Many equipment manufacturers publish specific flow and supply pressure requirements to achieve turbulent flow. Advantage temperature control units, portable and central liquid chillers, and pump tank stations are designed to generate turbulent flow.

Supply pumps should be designed for the flow and pressure requirements of the process. However, actual flow and velocity will depend on proper design and installation of the system. Supply and return plumbing should be designed for minimum restriction to flow, including minimum use of elbows and oversized quick connects. Full size pipes and hoses are recommended to avoid restricting the flow.

Turbulent flow also will extend the useful life of the process tooling by slowing the buildup of precipitates on the heat transfer surface.

Reynolds' experiment

The existence of laminar and turbulent flow is most easily visualized by the experiments of Reynold's. His experiments are shown in the Fig. 2.5. Water was allowed to flow at steady state through a transparent pipe with the flow rate controlled by a valve at the end of a pipe. A fine, steady stream of dyed water was introduced from a fine jet as shown and its flow pattern was observed.

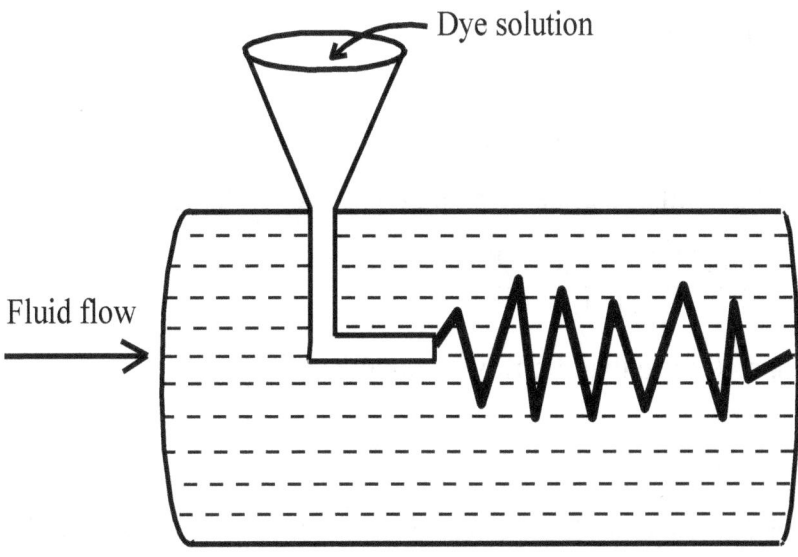

Fig. 2.5: Reynolds' apparatus

At low rates of water flow, the dye pattern was regular and formed a single line or stream similar to a thread, as shown in Fig.2.5. There was no lateral mixing of the fluid and it flowed in streamlines down the tube. This type of flow is called laminar or viscous flow.

When a laminar flow changes into a turbulent flow, it does not change abruptly but it has got some transition period between the two types of flows. This type of flow is called transient flow.

As the velocity was increased, it was found out that at a definite velocity, the thread of dye became dispersed and the pattern was very erratic. This type of flow is known as turbulent flow. The velocity at which the flow changes is known as the critical velocity.

Example 2.3: Flow regime of milk

1. Milk is flowing at 0.12 m³ min⁻¹ in a 2.5 cm diameter pipe. If the temperature of the milk is 21°C, is the flow turbulent or streamline? Viscosity and density of milk at 21°C is 2.1 cP and 1029 kg m⁻³ respectively.

Solⁿ:
Given,
Viscosity of milk at 21°C (μ) = 2.1 cP = 2.10 x 10⁻³ N s m⁻²
Density of milk at 21°C (ρ) = 1029 kg m⁻³.
Diameter of pipe (D) = 0.025 m.

Cross-sectional area of pipe (A) = $\frac{\pi}{4} D^2 = \frac{\pi}{4} \times 0.025^2$ = 4.9 x 10⁻⁴ m²

Rate of flow (Q*) = 0.12 m³ min⁻¹ = 2 × 10⁻³ m³ s⁻¹

So velocity of flow (v) = $\frac{Q^*}{A} = \frac{2 \times 10^{-3}}{4.9 \times 10^{-4}}$ = 4.1 m s⁻¹,

and so, Reynold's number can be calculated using equation 2.5 as;

$$(Re) = \frac{D v \rho}{\mu} = \frac{0.025 \times 4.1 \times 1029}{2.1 \times 10^{-3}} = 50{,}230$$

and, this is greater than 4000 so that the flow is *turbulent*.

Exercises on flow regimes

1. Whole milk at 293K having a density of 1030 kg/m³ and viscosity of 2.12 cp is flowing at the rate of 0.605 kg/s in a glass pipe having a diameter of 63.5 mm.
a. Calculate the Reynold's number. Is this turbulent flow?
b. Calculate the flow rate needed in m³/s for a Reynold's number of 2100 and velocity in m/s.
[**Ans:** 5723 and Turbulent]

2. An oil is being pumped inside a 10mm diameter pipe at a Reynold's number of 2100. The oil density is 855 kg/m³ and the viscosity is 2.1×10⁻² Pa.s.
a. What is the velocity in the pipe?
b. It is desired to maintain the same N_{Re} of 2100 and the same velocity as in part (a) using a second fluid with a density of 925 kg/m³ and a viscosity of 1.5×10⁻² Pa.s. What pipe diameter should be used? [**Ans:** 5.158 m/s and 0.0061m]

3. The flow rate of chocolate syrup is $4 m^3 s^{-1}$. And the cylindrical pipe in which the fluid flows has a diameter of 6 cm. Determine the Reynold's number for the fluid which has a density of 1268 kg/m^3 and a viscosity of 17 Pas. Indicate whether the flow regime is laminar or turbulent. [**Ans: 6332 and Turbulent**]

4. A fluid flows turbulently, at a rate of $3.5 m^3/s$ through a cylindrical pipe. The pipe has a cross-sectional area of 0.5 m^2 and a fluid has a viscosity of 1.2 N/sm^2 with Reynolds number of 6283. What is the density of the fluid? Assume the fluid is incompressible and the flow is steady.

5. Calculate the Reynold's number for water flowing at $5 m^3/h$ in a tube with 2 inch inside diameter if the viscosity and density of water are 1cp and 0.998g/ml respectively. [**Ans: 34770 & turbulent**]

6. Find the Reynolds number if a fluid of viscosity 0.4 Ns/m^2 and relative density of 900 Kg/m^3 through a 20 mm pipe with a Velocity of 2.5 m/s? [**Ans: 112.5**]

7. Calculate the Reynolds number if a fluid flows through a diameter of 80 mm with velocity 5 m/s having density of 1400 Kg/m^3 and having viscosity of 0.9 Kg/ms. [**Ans: 622.22**]

8. The 20 m long steel pipe below has a 25mm inner diameter. It carries water at a rate of 4.5 m^3/h. The density of water is 1000 kg/m^3, and water has an absolute viscosity of 1.00×10^{-3} Pas. Would the flow in the pipe be considered laminar, in transition or turbulent? [**Ans: 64000, Turbulent**]

Flow measuring equipments

The most common principals for fluid flow metering are differential pressure flowmeters, velocity flowmeters, positive displacement flowmeters, mass flowmeters and open channel flowmeters.

Flow meters are referred to by many names, such as flow gauge, flow indicator, liquid meter, etc. depending on the particular industry; however the function, to measure flow, remains the same.

In a differential pressure drop flowmeters, the flow is calculated by measuring the pressure drop over an obstructions inserted in the flow. The differential pressure flowmeter is based on the Bernoulli's equation, where the pressure drop and the further measured signal is a function of the square flow speed. Common types of differential pressure flowmeters are orifice plates, venturi tubes, pitot- tubes and variable area - rotameters.

Venturimeter

Another method of using pressure differentials to measure fluid flow rates is used in Venturi and orifice meters. Venturimeter is the most widely used device to measure the discharge through the pipe. A venturi is a converging-diverging nozzle of circular cross-section. If flow is constricted, there is a rise in velocity and a fall in static pressure in accordance with Bernoulli's equation.

Venturimeter consists of a short converging part, throat and a diverging part. A gradual constriction has been interposed in a pipe decreasing the area of flow from A_1 to A_2. Let the fluid is assumed to be incompressible and the respective velocities and static pressures are v_1 and v_2, and P_1 and P_2 as shown in Fig. 2.6.

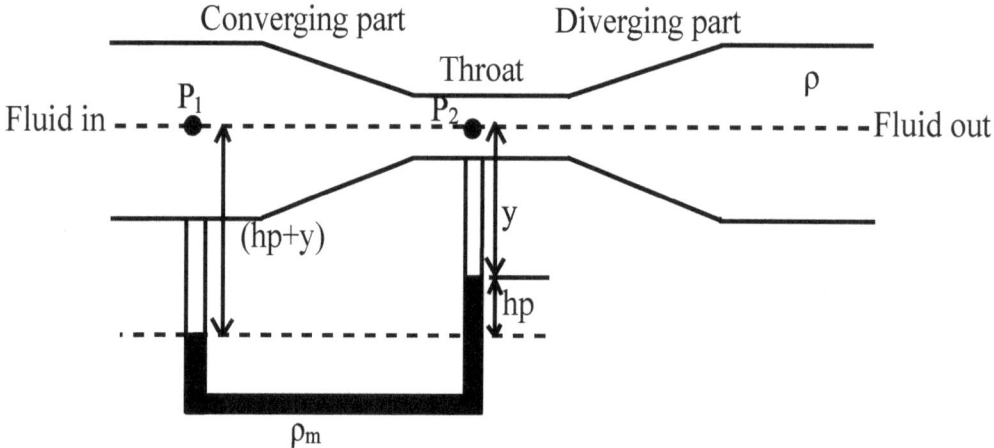

Fig. 2.6: Construction of venturimeter

1. Use Bernoulli's equation to solve for energy balance at two points:

   Total energy at point1 = Total energy at point2

   $$\text{Or, } P_1 + \frac{1}{2}\rho v_1^2 + \rho g h_1 = P_2 + \frac{1}{2}\rho v_2^2 + \rho g h_2$$

   Since, Venturimeter is at same level of base i.e. $h_1 = h_2$

   Hence,

   $$v_2^2 - v_1^2 = \frac{2(P_1 - P_2)}{\rho} \quad \ldots\ldots\ldots\ldots (2.6)$$

2. Use Continuity equation to solve for flow rates between two points:

   $$A_1 v_1 = A_2 v_2$$

   $$\text{Or, } v_2^2 = \left(\frac{A_1}{A_2}\right) v_1^2 \ldots\ldots\ldots (2.7)$$

Putting the value of v_2^2 in equation 2.6;

$$v_1^2 = \frac{2(P_1 - P_2)}{\rho\left\{\left(\frac{A_1}{A_2}\right)^2 - 1\right\}} \qquad \ldots\ldots\ldots (2.8)$$

By joining the two sections of a pipe to a U-manometer, as shown in the Fig.2.6, the differential pressure, $(P_2 - P_1)$ can be measured directly.

Solve for (P_1-P_2) using a differential manometer:

$P_1 + (h_p + y)\rho g = P_2 + h_p \rho_m g + y\rho g$

$P_1 - P_2 = h_p g (\rho_m - \rho)$ .... (2.9)

Calculate $P_1 - P_2$ and place the value in equation 2.8:

$$v_1^2 = \frac{2\Delta P}{\rho\left\{\left(\frac{A_1}{A_2}\right)^2 - 1\right\}}$$

Or, $\boxed{v_1 = \sqrt{\dfrac{2\Delta P}{\rho\left\{\left(\frac{A_1}{A_2}\right)^2 - 1\right\}}}}$ .... (2.10)

Theoritical flow rates

1. Volummetric flow rate is :

$$Q^* = Av = A_1 \sqrt{\frac{2\Delta P}{\rho\left\{\left(\frac{A_1}{A_2}\right)^2 - 1\right\}}} \qquad \ldots\ldots\ldots (2.11)$$

2. Mass flow rate is :

$$m^* = \rho Av = \rho A_1 \sqrt{\frac{2\Delta P}{\rho\left\{\left(\frac{A_1}{A_2}\right)^2 - 1\right\}}} \qquad \ldots\ldots\ldots (2.12)$$

Discharge coefficient (C_D)

Actual flowrate will always be less than the theoretical flowrate values due to friction and other losses.

$$C_D = \frac{\text{Actual flowrate}}{\text{Theoretical flowrate}}$$

Typical value of C_D for venturimeter is 0.98 while that for orificemeter is 0.65.

Practical flowrates using C_D

1. Volumetric flow rate is:

$$Q^* = C_D Av = C_D A_1 \sqrt{\frac{2\Delta P}{\rho\left\{\left(\frac{A_1}{A_2}\right)^2 - 1\right\}}} \quad \ldots\ldots(2.13)$$

2. Mass flow rate is:

$$m^* = C_D \rho Av = C_D \rho A_1 \sqrt{\frac{2\Delta P}{\rho\left\{\left(\frac{A_1}{A_2}\right)^2 - 1\right\}}} \quad \ldots\ldots(2.14)$$

Example 2.4: Flowrate in venturimeter

1. A venturimeter is 50mm bore diameter at inlet and 10mm bore diameter at the throat. Oil of density 900kg/m³ flows through it and a differential pressure head of 80mm is produced. Given C_D=0.92, determine the flow rate in kg/s.

Sol^n:

Given,

Inlet diameter of the tube (d_1) = 50 mm = 50×10^{-3} m

Diameter at the throat (d_2) = 10 mm = 10×10^{-3} m

Density of oil (ρ_{oil}) = 900 Kgm^{-3}

Differential pressure head (Δh) = 80 mm = 80×10^{-3}

Discharge coefficient of meter (C_D) = 0.92

Mass flowrate (m^*) =?

Cross-sectional area of tube at inlet (A_1) = $\frac{\pi d_1^2}{4}$ = 0.0019625 m²

Cross-sectional area at the throat $(A_2) = \dfrac{\pi d_2^2}{4} = 0.0000785 \text{ m}^2$

Differential pressure $(\Delta P) = \Delta h \times \rho_{oil} \times g$
$= 80 \times 10^{-3} \times 900 \times 9.8 = 705.6$ Pa.

Therefore, using equation 2.10;

Inlet velocity $(v_1) = \sqrt{\dfrac{2\Delta P}{\rho\left\{\left(\dfrac{A_1}{A_2}\right)^2 - 1\right\}}} = \sqrt{\dfrac{2 \times 705.6}{900\left\{\left(\dfrac{0.0019625}{0.0000785}\right)^2 - 1\right\}}} = 0.050127 \text{ ms}^{-1}$

Hence, using equation 2.14;

m* = $C_D \rho A_1 v_1$ = 0.92 × 900 × 0.0019625 × 0.050127 = 0.08145 Kg/s.

Hence, the required flow rate is found to be 0.08145 Kg/s.

Exercises on venturimeter

1. A venturimeter with a 15 cm diameter at inlet and 10 cm throat is laid with its axis horizontal and is used for measuring the flow of oil of sp.gr 0.9. The oil-mercury differential manometer shows a gauge difference of 20 cm. Assume coefficient of the meter as 0.98, calculate the discharge. [**Ans:** 0.063837 m³/s]

2. A horizontal venturimeter 160mm× 80mm is used to measure the flow of an oil of sp.gr 0.8. Determine the deflection of the oil-mercury gauge, if the discharge of the oil is 50L/s. [**Ans:** 0.296m]

3. Calculate the differential pressure expected from a Venturimeter when the flow rate is 2 dm³/s of water. The area ratio is 4 and C_D is 0.94. The inlet cross-sectional area is 900 mm². [**Ans:** 41.916 kPa]

4. A venturimeter is 60mm bore diameter at inlet and 20mm bore diameter at the throat. Water of density 1000kg/m³ flows through it and a differential pressure head of 150mm is produced. Given C_D=0.95, determine the flow rate in dm³/s. [Hint:1000m³=1dm³] [**Ans:** 0.515dm³/s]

5. Calculate the mass flow rate of water through a venturimeter when the differential pressure is 980 Pa. Given C_D is 0.93, the area ratio is 5 and the inlet cross sectional area is 1000mm². [**Ans:** 0.266 kg/s]

6. The air of velocity 15 m/s and of density 1.3 kg/m³ is entering the Venturi tube (placed in the horizontal position) from the right. The radius of the wide part of the tube is 1.0 cm; the radius of the thin part of the tube is 0.5 cm. The tube of shape U connecting wide and thin part of the main tube is filled with the mercury of the density 13,600 kg/m³. Determine what height difference will be stabilized between the surfaces of the mercury in U-tube. [**Ans:** 1.6 cm]

7. A Horizontal Venturi meter with d_1 = 20 cm, and d_2 = 10 cm, is used to measure the flow rate of oil of sp.gr. = 0.8, the discharge through venture meter is 60 lit/s. find the reading of (oil-Hg) differential. Take C_D = 0.98. [**Ans:** 18.15 cm Hg]

8. A horizontal Venturi meter is used to measure the flow rate of water through the piping system of 20 cm I.D, where the diameter of throat in the meter is d_2 = 10 cm. The pressure at inlet is 17.658 N/cm² gauge and the vacuum pressure of 35 cm Hg at throat. Find the discharge of water. Take C_D = 0.98. [**Ans:** 0.168 m³/s]

9. A Venturi meter is to be fitted to a 25 cm diameter pipe, in which the maximum flow is 7200 lit/min and the pressure head is 6 m of water. What is the maximum diameter of throat, so that there is non-negative head on it? [Hint h_2=0] [**Ans:** 11.72 cm]

10. A Venturi meter with a 15 cm I.D. at inlet and 10 cm I.D. at throat is laid with its axis horizontal and is used for measuring the flow of oil of sp.gr. = 0.9. The oil-mercury differential manometer shows a gauge difference of 20 cm. If C_D = 0.98, calculate the discharge of oil. [**Ans:** 0.06393 m³/s]

11. A Venturi meter has an area ratio (9:1), the larger diameter being 30 cm. During the flow the recorded pressure head in larger section is 6.5 m and that at throat 4.25 m. If C_D = 0.99, compute the discharge through the meter. [**Ans:** 0.052 m³/s]

12. The air of velocity 15 m/s and of density 1.3 kg/m³ is entering the venturi tube from the right. The radius of the wide part of the tube is 1.0 cm; the radius of the thin part of the tube is 0.5 cm. The tube of shape U connecting wide and thin part of the main tube is filled with the mercury of the density 13,600 kg/m³. Determine what height difference will be stabilized between the surfaces of the mercury in U-tube. [**Ans:** 1.6 cm]

Orificemeter

An orifice meter or orifice plate is a flat plate having a central hole that is placed across the flow of a liquid, usually between flanges in a pipeline. The pressure difference generated by the flow velocity through the hole enables the flow quantity to be measured.

The orifice meter operates on the same principle as the Venturi meter, constricting the flow and measuring the corresponding static pressure drop as shown in Fig. 2.7 below.

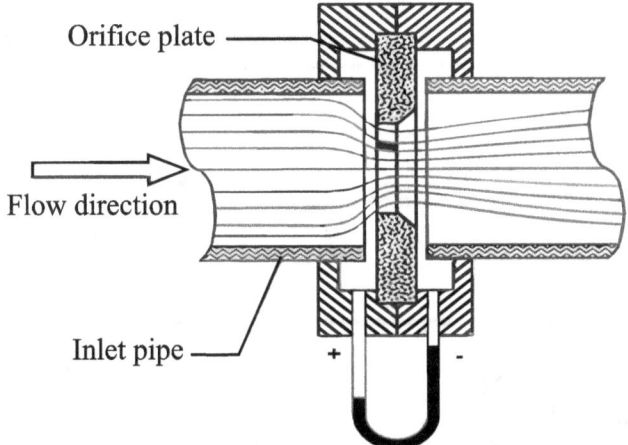

Fig. 2.7: Orificemeter

Instead of a tapered tube, a plate with a hole in the centre is inserted in the pipe to cause the pressure difference. The same equations hold as for the Venturi meter; but in the case of the orifice meter the coefficient, called the orifice discharge coefficient, is smaller. Orifices have much greater pressure losses than Venturi meters, but they are easier to construct and to insert in pipes.

Exercises on orificemeter

1. An orifice meter consisting of 10 cm diameter orifice in a 25 cm diameter pipe has C_D of 0.65. The pipe delivers oil of sp.gr. = 0.8. The pressure difference on the two sides of the orifice plate is measured by mercury oil differential manometer. If the differential gauge is 80 cm Hg, find the rate of flow. [**Ans:** 0.08196 m^3/s]

2. Calculate the flow rate of water through an orifice meter with an area ratio of 4. Given C_D is 0.62, the pipe area is 900mm^2 and the differential pressure is 586 Pa. [**Ans:** 0.156 dm^3/s]

3. An orificemeter is used to measure liquid flow rate of 7500 litres per minute. The difference in pressure across the orificemeter is equivalent to 8 m of the flowing liquid. The pipe diameter is 19 cm. Calculate the throat diameter of the orificemeter. [**Ans:** 10 cm]

Pitot tube

The Pitot tube (named after Henri Pitot in 1732) measures a fluid velocity by converting the kinetic energy of the flow into potential energy. The conversion takes place at the stagnation point, located at the Pitot tube entrance. A pressure higher than the free-stream (i.e. dynamic) pressure results from the kinematic to potential conversion. This "static" pressure is measured by comparing it to the flow's dynamic pressure with a differential manometer as shown in Fig. 2.8.

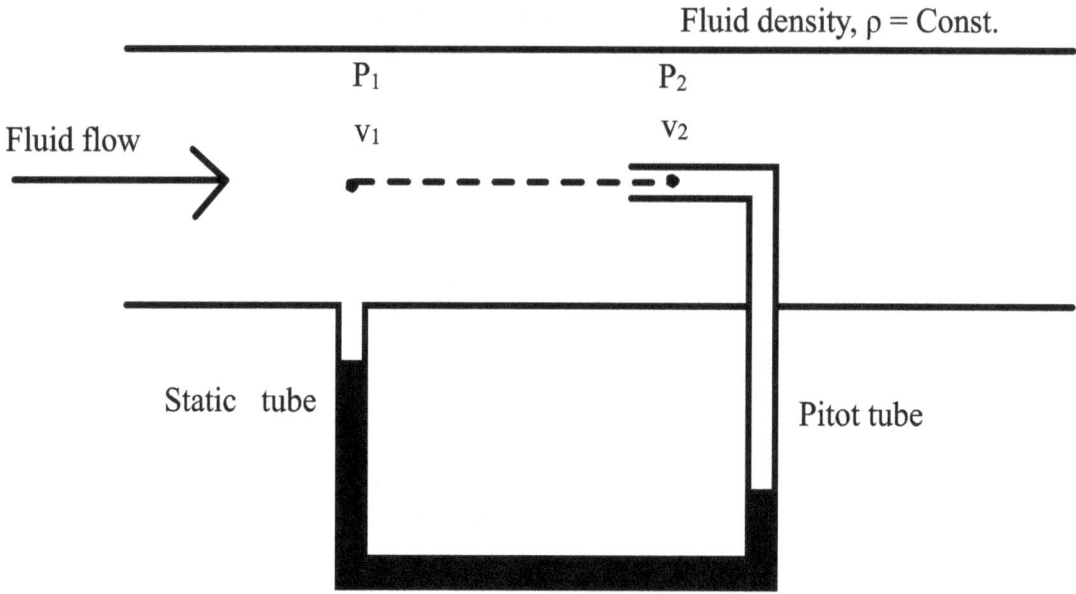

Fig. 2.8: Construction of pitot tube

Using Bernoulli's equation between two points where $v_2 = 0$ and $h_1 = h_2$.

$$P_1 + \frac{1}{2}\rho v_1^2 = P_2 + \frac{1}{2}\rho v_2^2$$

Or, $P_2 - P_1 = \frac{1}{2}\rho v_1^2$

Or, $v_1 = \sqrt{\dfrac{2(P_2 - P_1)}{\rho}}$ .... (2.15)

$(P_2 - P_1)$ value can be found using differential manometer.

i.e. $(P_2 - P_1) = (\rho_m - \rho)\, g h_p$.

Hence, $\boxed{\text{Volumetric flow rate } (Q) = A_1 \cdot \sqrt{\dfrac{2(P_2 - P_1)}{\rho}}}$ .... (2.16)

Example 2.5: Velocity of oil in pitot tube

1. Find the local velocity of the flow of an oil of sp.gr. =0.8 through a pipe, when the difference of mercury level in differential U-tube manometer connected to the two tapping of the Pitot tube is 10 cm Hg. Take Cp = 0.98.

Sol^n:

Given,

Density of mercury (ρ_m) = 13,600 Kg/m^3
Density of oil (ρ) = 0.8 × 100 = 800 Kg/m^3
Difference of mercury level (h_p) = 10 cm = 0.1 m.
Discharge coefficient of pitot tube (C_p) = 0.98
Velocity of oil (v) = ?

For differential manometer, pressure difference can be found out by:

$$P_2 - P_1 = (\rho_m - \rho) g h_p$$
$$= (13,600 - 800) \times 9.8 \times 0.1$$
$$= 12,554 \text{ Pa}$$

Therefore, using equation 2.15;

$$v = C_p \sqrt{\frac{2(P_2 - P_1)}{\rho}} = 0.98 \times \sqrt{\frac{2 \times 12554}{800}} = 5.488 \text{ m/s}.$$

Hence, the velocity of oil is found to be 5.488 m/s.

Exercises on pitot tube

1. A Pitot tube is placed at a center of a 30 cm I.D. pipe line has one orifice pointing upstream and other perpendicular to it. The mean velocity in the pipe is 0.84 of the center velocity (i.e. u/u$_x$ =0.94). Find the discharge through the pipe if: i. The fluid flow through the pipe is water and the pressure difference between orifice is 6 cm H$_2$O. ii. The fluid flow through the pipe is oil of sp.gr. 0.78 and the reading manometer is 6 cm H$_2$O. Take Cp = 0.98. [**Ans:** 0.475 m/s, 0.0335 m^3/s]

2. A Pitot tube is inserted in the pipe of 30 cm I.D. The static pressure head is 10 cm Hg vacuum, and the stagnation pressure at center of the pipe is 0.981 N/cm^2 gauge. Calculate the discharge of water through the pipe if u/u$_{max}$ = 0.85. Take Cp = 0.98. [**Ans:** 5.67 m/s, 0.4 m^3/s]

3. Find the local velocity of the flow of an oil of sp.gr. =0.8 through a pipe, when the difference of mercury level in differential U-tube manometer connected to the two tapping of the Pitot tube is 10 cm Hg. Take Cp = 0.98. [**Ans:** 1.3m/s]

Rotameter

Rotameter is a variable area meter, i.e. it measures the area of flow so as to produce a constant head differential. Therefore, rotameters are also called area meters. The construction of rotameter is shown in Fig. 2.9.

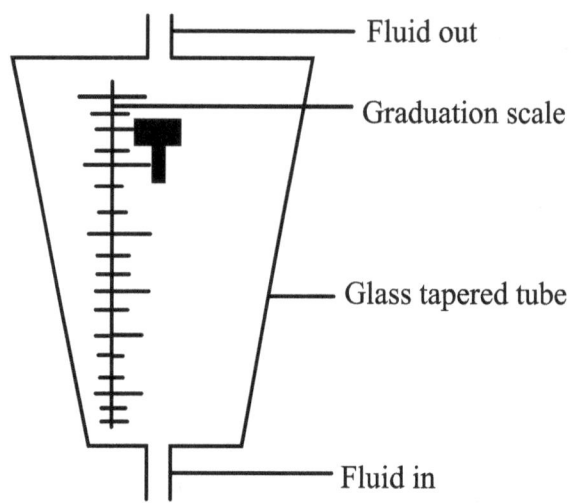

Fig. 2.9: Construction of rotameter

It consists of a vertical, tampered and transparent tubes in which a plummet is placed. During the fluid flow through the tube, the plummet rises and falls because of variation in the flow. As a result, the area of the annular space between the plummet and the tube varies. The head losses across the annulus is equal to the weight of the plummet. The upper edge of the plummet is used as a index to note the reading on the tapered tube. This value indicates the flowrate of the fluid.

Uses

1. Extensively used in chemical industries, such as bulk drugs.
2. The supply of air in fermenters is controlled through rotameters.

Advantages

1. Requires no external power or fuel. It uses only the inherent properties of the fluid, along with gravity, to measure the flow rate.
2. Operator has a direct visual index of flow reading.
3. It does not require the condition that straight pipes should run before and after the meter.

Viscosity

Viscosity is a property of the fluid which opposes the relative motion between the two surfaces of the fluid in a fluid that are moving at different velocities. When the fluid is forced through a tube, the particles which compose the fluid generally move more quickly near the tube's axis and more slowly near its walls; therefore some stress(such as a pressure difference between the two ends of the tube) is needed to overcome the friction between particle layers to keep the fluid moving. For a given velocity pattern, the stress required is proportional to the fluid's viscosity.

Viscosity is that property of a fluid that gives rise to forces that resist the relative movement of adjacent layers in the fluid. Viscous forces are of the same character as shear forces in solids and they arise from forces that exist between the molecules.

If two parallel plane elements in a fluid are moving relative to one another, it is found that a steady force must be applied to maintain a constant relative speed. This force is called the viscous drag because it arises from the action of viscous forces. Consider the system shown in Fig. 2.10.

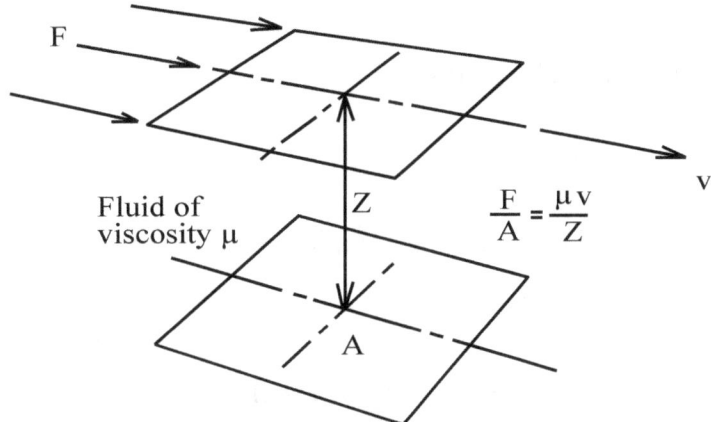

Fig. 2.10: Viscous forces in a fluid

If the plane elements are at a distance Z apart, and if their relative velocity is v, then the force F required to maintain the motion has been found, experimentally, to be proportional to v and inversely proportional to Z for many fluids. The coefficient of proportionality is called the viscosity of the fluid, and it is denoted by the symbol μ (mu).

From the definition of viscosity we can write:

$$\frac{F}{A} = \frac{\mu v}{Z} \quad \ldots\ldots\ldots (2.17)$$

where F is the force applied, A is the area over which force is applied, Z is the distance between planes, v is the velocity of the planes relative to one another, and μ is the viscosity.

Newtonian and non-newtonian fluids

Newtonian fluids are those having a constant viscosity dependent on temperature but independent of the applied shear rate. One can also say that Newtonian fluids have direct proportionality between shear stress and shear rate in laminar flow. Materials which cannot be defined by a single viscosity value at a specified temperature are called *non-Newtonian*. The viscosity of these materials must always be stated together with a corresponding temperature and shear rate. If the shear rate is changed, the viscosity will also change.

From the fundamental definition of viscosity in equation 2.17, we can write:

$$\frac{F}{A} = \frac{\mu v}{Z} = \mu \left(\frac{dv}{dz}\right) = \tau \quad \ldots\ldots\ldots \text{(2.18)}$$

where τ (tau) is called the shear stress in the fluid. This is an equation originally proposed by Newton and which is obeyed by fluids such as water. However, for many of the actual fluids encountered in the food industry, measurements show deviations from this simple relationship, and lead towards a more general equation:

$$\tau = k \left(\frac{dv}{dz}\right)^n \quad \ldots\ldots\ldots \text{(2.19)}$$

which can be called the power-law equation, and where k is a constant of proportionality.

Where n = 1 the fluids are called Newtonian because they conform to Newton's equation 2.19 and k = μ; and all other fluids may therefore be called non-Newtonian. Non-Newtonian fluids are varied and are studied under the heading of rheology, which is a substantial subject in itself and the subject of many books. Broadly, the non-Newtonian fluids can be divided into:

(1) <u>Those in which *n* < 1</u>. As shown in Fig. 2.11, these produce a concave downward curve and for them the viscosity is apparently high under low shear forces decreasing as the shear force increases. Such fluids are called pseudoplastic, an example being tomato puree. In more extreme cases where the shear forces are low there may be no flow at all until a yield stress is reached after which flow occurs, and these fluids are called thixotropic.

(2) <u>Those in which *n* > 1</u>. With a low apparent viscosity under low shear stresses, they become more viscous as the shear rate rises. This is called dilatancy and examples are gritty slurries such as crystallized sugar solutions. Again there is a more extreme condition with a zero apparent viscosity under low shear and such materials are called rheopectic. Bingham fluids have to exceed a particular shear stress level (a yield stress) before they start to move.

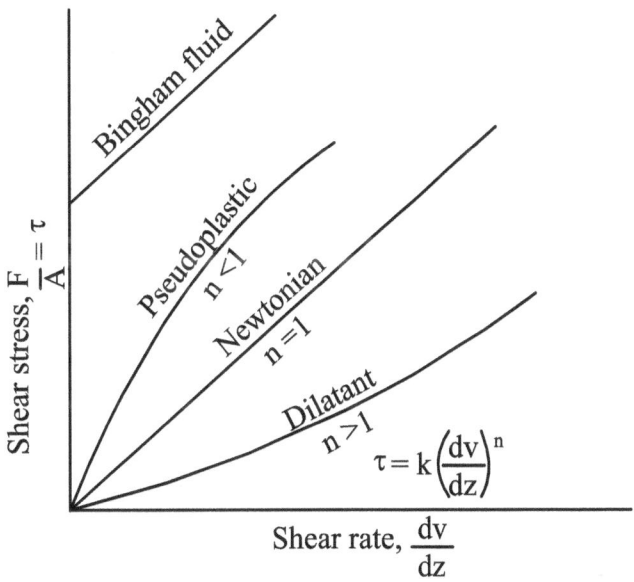

Fig. 2.11: Shear stress/shear rate relationships in liquids

In many instances in practice non-Newtonian characteristics are important, and they become obvious when materials that it is thought ought to pump quite easily just do not. They get stuck in the pipes, or overload the pumps, or need specially designed fittings before they can be moved. Sometimes it is sufficient just to be aware of the general classes of behavior of such materials. In other cases it may be necessary to determine experimentally the rheological properties of the material so that equipment and processes can be adequately designed.

Fanning equation

Consider an energy balance over a differential length, dL, of a straight horizontal pipe of diameter D, as in Fig. 2.12 below.

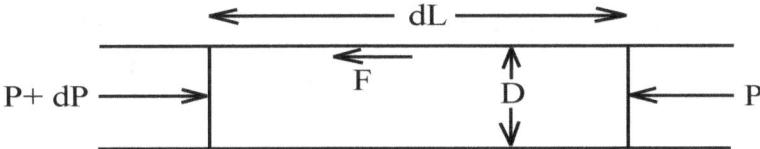

Fig. 2.12: Energy balance over a length of pipe

Consider the equilibrium of the element of fluid in the length dL. The total force required to overcome friction drag must be supplied by a pressure force giving rise to a pressure drop dP along the length dL.

The pressure drop force is:

$$dP \times \text{Area of pipe} = dP \times \frac{\pi D^2}{4} \quad \ldots\ldots\ldots \text{(2.20)}$$

The friction force is (force/unit area) x wall area of pipe:

$$= \frac{F}{A} \times \pi D \times dL = \left(\frac{f\rho v^2}{2}\right) \times \pi D \times dL \dots\dots\dots (2.21)$$

At equilibrium state, pressure drop force and friction force must be equal.

i.e. $\left(\frac{\pi D^2}{4}\right) dP = \left(\frac{f\rho v^2}{2}\right) \times \pi D \times dL,$

Therefore, $dP = 4\left(\frac{f\rho v^2}{2}\right) \times \frac{dL}{D}\dots\dots\dots(2.22)$

Integrating between L_1 and L_2, in which interval P goes from P_1 to P_2 we have:

$$\int_{P_1}^{P_2} dP = \int_{L_1}^{L_2}\left[4\left(\frac{f\rho v^2}{2}\right) \times \frac{dL}{D}\right]$$

Or, $P_1 - P_2 = \left(\frac{4f\rho v^2}{2}\right)\left(\frac{L_1 - L_2}{D}\right)$

i.e. $\Delta P_f = \left(\frac{4f\rho v^2}{2}\right) \times \left(\frac{L}{D}\right)\dots\dots\dots (2.23)$

Hence, $\boxed{E_f = \frac{\Delta P_f}{\rho} = \frac{2fv^2 L}{D}}\dots\dots\dots (2.24)$

where $L = L_1 - L_2$ = length of pipe in which the pressure drop, $\Delta P_f = P_1 - P_2$ is the frictional pressure drop, and E_f is the frictional loss of energy.

Equation 2.24 is an important equation; it is known as the Fanning equation, or sometimes the D'Arcy or the Fanning-D'Arcy equation. It is used to calculate the pressure drop that occurs when liquids flow in pipes. The factor f in equation 2.24 depends upon the Reynolds number for the flow, and upon the roughness of the pipe.

Friction factors
Fanning friction factor

It is a dimensionless number used in fluid flow calculations. It is related to the shear stress at the wall as:

$$\tau = \frac{F}{A} = \left(\frac{f\rho v^2}{2}\right) \dots\dots\dots (2.25)$$

For laminar flow:

f = 16/ Re, for round tubes, &

f = 14.227/ Re, for square channel

Darcy friction factor

It is a dimensionless quantity which describes friction losses in pipe flow as well as open flow. It is 4 times larger than the fanning friction factor.

For laminar flow:

f = 64/ Re, for round tubes, &

f = 56.9/ Re, for square channel

Example 2.6: Pressure drop of olive oil

1. Calculate the pressure drop along 170 m of 5 cm diameter horizontal steel pipe through which olive oil at 20°C is flowing at the rate of 0.1 m³ min⁻¹. The viscosity of olive oil at 20°C is 84 x 10⁻³ Ns m⁻² and density = 910 kg m⁻³.

Sol^n:

Given,

Diameter of pipe (D) = 0.05 m,

Area of cross-section (A) = $\left(\frac{\pi}{4}\right) D^2$

$= \left(\frac{\pi}{4}\right) \times (0.05)^2 = 1.96 \times 10^{-3}$ m²

Velocity of oil (v) = $\left(\dfrac{0.1 \times \frac{1}{60}}{1.96 \times 10^{-3}}\right) = 0.85$ m s⁻¹,

Now, using equation 2.5;

Reynolds' number (Re) = $\dfrac{Dv\rho}{\mu} = \dfrac{0.05 \times 0.85 \times 910}{84 \times 10^{-3}} = 460$

For streamline flow, f = $\dfrac{16}{Re} = \dfrac{16}{460} = 0.03$

And so the pressure drop in 170 m, from equation 2.23;

$$\Delta P_f = \left(\dfrac{4f\rho v^2}{2}\right) \times \left(\dfrac{L}{D}\right) = \left\{\dfrac{4 \times 0.03 \times 910 \times (0.85)^2}{2}\right\} \times \left(\dfrac{170}{0.05}\right) = 1.34 \times 10^5 \text{ Pa} = 134 \text{ KPa}.$$

Hence, the required pressure drop is found to be 134 KPa.

Moody's diagram

The friction factor 'f' depends upon the Reynolds number for the flow, and upon the roughness of the pipe. If the Reynolds number and the roughness factor are known, then f can be read off from the graph shown in Fig. 2.13 below.

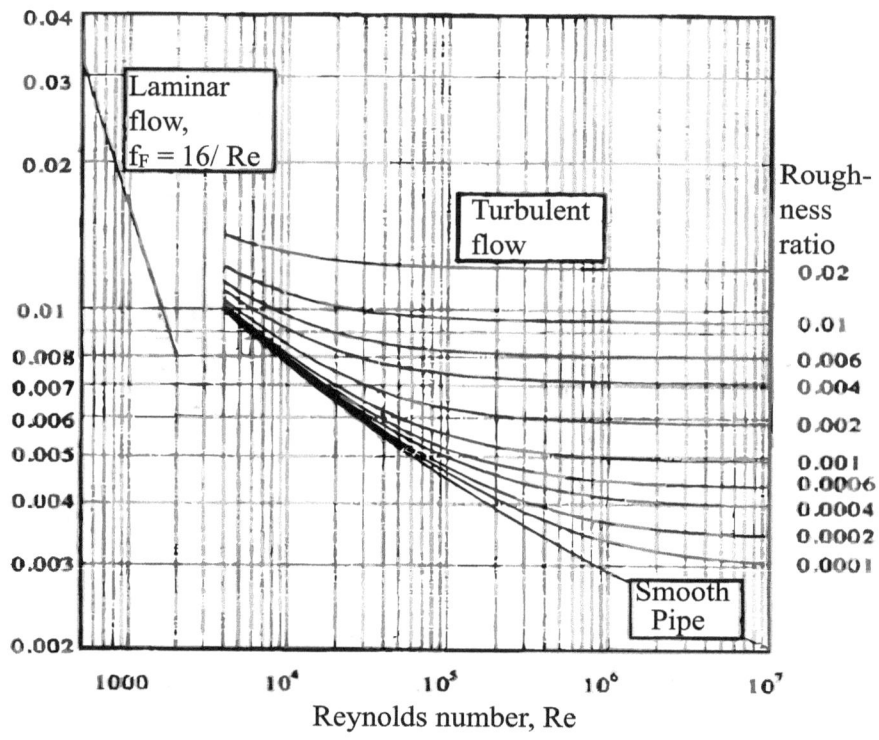

Fig. 2.13: Moody's diagram

Example:
If, N_{Re}=20000, Roughness =0.004, then friction factor, f =0.0078

Energy losses in bends and fittings

When the direction of flow is altered or distorted, as when the fluid is flowing round bends in the pipe or through fittings of varying cross-section, energy losses occur which are not recovered. This energy is dissipated in eddies and additional turbulence and finally lost in the form of heat. However, this energy must be supplied if the fluid is to be maintained in motion, in the same way, as energy must be provided to overcome friction. Losses in fittings have been found, as might be expected, to be proportional to the velocity head of the fluid flowing. In some cases the magnitude of the losses can be calculated but more often they are best found from tabulated values based largely on experimental results. The energy loss is expressed in the general form,

$$E_f = \frac{kv^2}{2} \quad \ldots\ldots\ldots \quad (2.26)$$

where k has to be found for the particular fitting. Values of this constant k for some fittings are given in Table 2.1.

Table 2.1: Friction loss factors in fittings

	k
• Valves, fully open:	
- gate	0.13
- globe	6.0
- angle	3.0
• Elbows:	
- 90° standard	0.74
- medium sweep	0.5
- long radius	0.25
- square	1.5
• Tee, used as elbow	1.5
• Tee, straight through	0.5
• Entrance, large tank to pipe:	
- sharp	0.5
- rounded	0.05

Energy is also lost at sudden changes in pipe cross-section.

i. <u>At sudden enlargement</u>

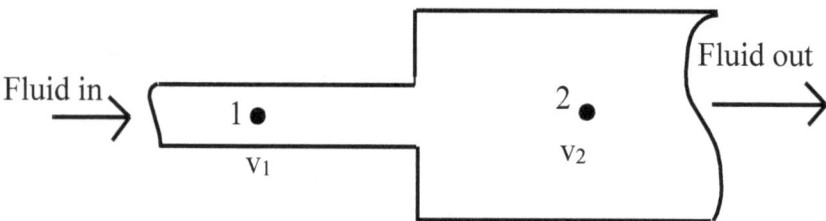

At a sudden enlargement the loss has been shown to be equal to:

$$E_f = \frac{(v_1 - v_2)^2}{2} \quad \ldots\ldots\ldots (2.27)$$

ii. <u>At sudden contraction</u>

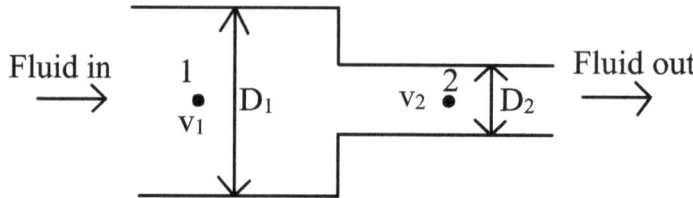

For a sudden contraction

$$E_f = \frac{k v_2^2}{2} \quad \ldots\ldots\ldots (2.28)$$

where v_1 is the velocity upstream of the change in section and v_2 is the velocity downstream of the change in pipe diameter from D_1 to D_2.

The coefficient k in equation 2.28 depends upon the ratio of the pipe diameters (D_2/D_1) as given in Table 2.2.

Table 2.2: Loss factors in contractions

D_2/D_1	0.1	0.3	0.5	0.7	0.9
k	0.36	0.31	0.22	0.11	0.02

Chapter 3: Heat transfer

Introduction

Heat transfer is an operation that occurs repeatedly in the food industry. Whether it is called cooking, baking, drying, sterilizing or freezing, heat transfer is part of the processing of almost every food. An understanding of the principles that govern heat transfer is essential to an understanding of food processing.

Heat transfer is a dynamic process in which heat is transferred spontaneously from one body to another cooler body. The rate of heat transfer depends upon the differences in temperature between the bodies, the greater the difference in temperature, the greater the rate of heat transfer.

Temperature difference between the source of heat and the receiver of heat is therefore the driving force in heat transfer. An increase in the temperature difference increases the driving force and therefore increases the rate of heat transfer. The heat passing from one body to another travels through some medium which in general offers resistance to the heat flow. Both these factors, the temperature difference and the resistance to heat flow, affect the rate of heat transfer. As with other rate processes, these factors are connected by the general equation:

$$\text{Rate of transfer} = \frac{\text{driving force}}{\text{resistance}}$$

<u>For heat transfer</u>

$$\text{Rate of heat transfer} = \frac{\text{temperature difference}}{\text{heat flow resistance of medium}}$$

Two substances must have different temperatures in order to transfer heat from one to the other. Heat always flows from the warmer substance to the colder. The heat flow is rapid when the temperature difference is great. During heat transfer, the difference in temperature is gradually reduced and the rate of transfer slows down, ceasing altogether when the temperatures are equalized.

During processing, temperatures may change and therefore the rate of heat transfer will change. This is called unsteady state heat transfer, in contrast to steady state heat transfer when the temperatures do not change. An example of unsteady state heat transfer is the heating and cooling of cans in a retort to sterilize the contents. Unsteady state heat transfer is more complex since an additional variable, time, enters into the rate equations.

Mechanism of heat transfer

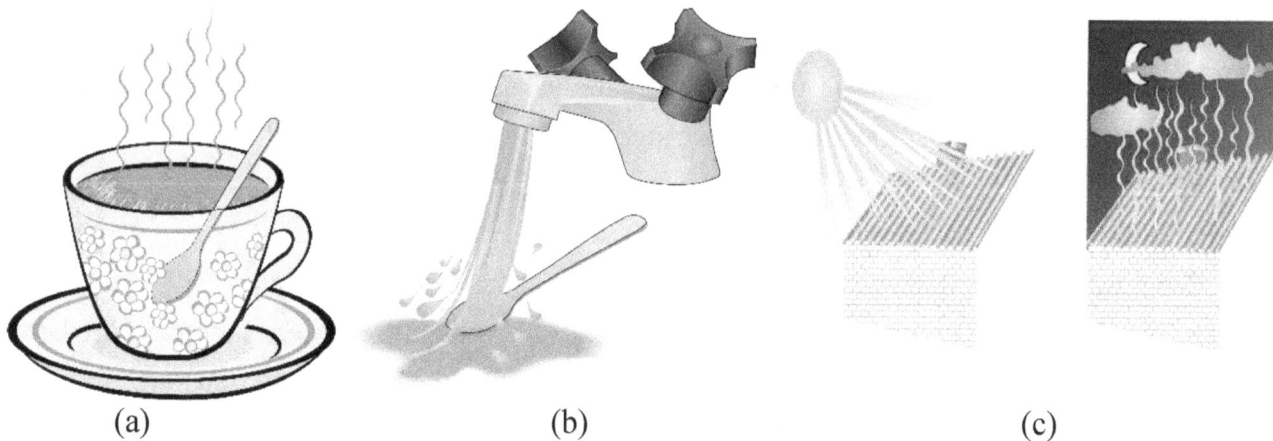

(a) (b) (c)

Fig. 3.1: Mechanism of heat transfer by (a) conduction (b) convection (c) radiation

Heat can be transferred in three ways: by conduction, by radiation and by convection.

In conduction, the molecular energy is directly exchanged, from the hotter to the cooler regions, the molecules with greater energy communicating some of this energy to neighbouring molecules with less energy. An example of conduction is the heat transfer through the solid walls of a refrigerated store. Eg., Fig. 3.1 (a) Heat is transferred from the bowl of the spoon to the handle.

Convection is the transfer of heat by the movement of groups of molecules in a fluid. The groups of molecules may be moved by either density changes or by forced motion of the fluid. An example of convection heating is cooking in a jacketed pan: without a stirrer, density changes cause heat transfer by natural convection; with a stirrer, the convection is forced. Eg., Fig. 3.1 (b) The spoon is rinsed in running cold water.

Radiation is the transfer of heat energy by electromagnetic waves, which transfer heat from one body to another, in the same way as electromagnetic light waves transfer light energy. An example of radiant heat transfer is when a foodstuff is passed below a bank of electric resistance heaters that are red-hot. Eg., Fig. 3.1 (c) A roof accumulates solar heat during the day and radiates the heat at night.

In general, heat is transferred in solids by conduction, in fluids by conduction and convection. Heat transfer by radiation occurs through open space, can often be neglected, and is most significant when temperature differences are substantial. In practice, the three types of heat transfer may occur together. For calculations it is often best to consider the mechanisms separately, and then to combine them where necessary.

Fourier's law of heat conduction

It states that "the rate of heat transfer through a uniform material is proportional to the area, the temperature drop and inversely proportional to the length of the path of flow".

$q_x \propto A \cdot \dfrac{dT}{dx}$

Or, $\dfrac{q_x}{A} = -k \cdot \dfrac{dT}{dx}$ .... (3.1)

where, q_x is the heat transfer rate in x- direction in watts (W), A is the cross- sectional area normal to the direction of flow of heat in m². T is the temperature in Kelvin, x is the distance in m and K is the thermal conductivity in W/ mK.

The quantity q_x/A is the heat flux in W/m². The quantity dT/dx is the temperature gradient in the x- direction. Negative sign indicates that the temperature decreases with positive heat flow.

Conduction through a flat slab or wall

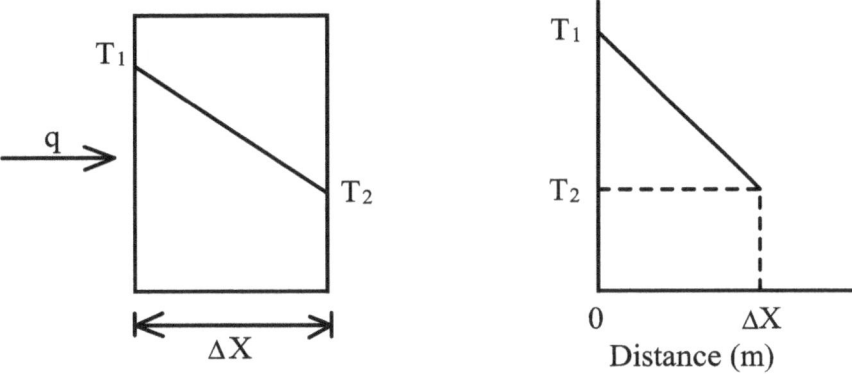

Fig. 3.2: Conduction through a flat slab

If a slab of material, as shown in Fig. 3.2, has two faces at different temperatures T_1 and T_2 heat will flow from the face at the higher temperature T_1 to the other face at the lower temperature T_2.

The rate of heat transfer is given by Fourier's equation 3.1;

$$\dfrac{dQ}{dt} = kA \dfrac{\Delta T}{\Delta x} = kA \dfrac{dT}{dx}$$

Integrating Fourier's law:

$$\dfrac{q}{A} = k \dfrac{[T_1 - T_2]}{[x_2 - x_1]}$$

$$q = kA \frac{[T_1 - T_2]}{\Delta x}$$

Or, $q = \dfrac{T_1 - T_2}{\dfrac{\Delta x}{KA}} = \dfrac{\text{Driving force}}{\text{Resistance}}$ .... (3.2)

Example 3.1: Heat transfer through a glass window

1. Calculate the rate of heat transfer through a glass window with 3m² surface area and 5mm thickness if the temperature on the two sides of the glass is 14°C and 15°C respectively and the thermal conductivity of the glass is 0.7 W/m°C. The system is at steady state.

Sol^n:
Given,
Temperature of glass on outer side (T_1) = 15°C
Temperature of glass on inner side (T_2) = 14°C
Area of glass window (A) = 3m².
Thickness of glass window (Δx) = 5mm = 0.005m
Thermal conductivity of the glass (k) = 0.7W/m°C
Rate of heat transfer (q) =?

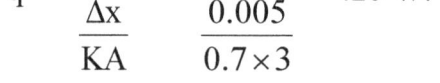

From equation 3.2, we have:

$$q = \frac{T_1 - T_2}{\dfrac{\Delta x}{KA}} = \frac{15 - 14}{\dfrac{0.005}{0.7 \times 3}} = 420 \text{ W}.$$

Hence, the rate of heat transfer through a glass window is 420 W.

Exercises on heat transfer through a single slab

1. A cork slab 10 cm thick has one face at -12°C and the other face at 21°C. If the mean thermal conductivity of cork in this temperature range is 0.042 J m^{-1} s^{-1} °C^{-1}, what is the rate of heat transfer through 1 m² of wall? [**Ans:** 13.9 J s^{-1}]

2. Calculate the heat loss per m² of surface area for an insulating wall composed of 25.4 mm thick fibre insulating board, where the inside temperature is 352.7 K and the outside temperature is 297.1 K. The thermal conductivity of the wall is 0.048 W/mK. [**Ans:** 105.07 W/m²]

3. A layer of fat 5 mm thick underneath the skin covers a part of a human body. If the temperature of the inner surface of the fat layer is 36.6°C and the body loses heat at a rate of 200 W/m^2, what will be the temperature at the surface of the skin? Assume that the thermal conductivity of fat is 0.2 W/m°C.

4. Asbestos layer of 10mm thickness (k = 0.116W/mK) is used as insulation over a boiler wall. Consider an area of 0.5m^2, find out the rate of heat flow as well as the heat flux over this area if the temperatures on either side of the insulation are 300°C and 30°C. [**Ans:** 3132 W/m^2 and 1566 W]

5. Determine the heat transfer rate per square meter of surface of a cork board, 3 cm thick, when a temperature difference of 75°C is applied across the board. Take the value of thermal conductivity (k) of cork board as -0.4 W/m°C. [**Ans:** 90 W]

6. A cold storage consists of a cubical chamber of dimension 2m x 2m x 2m, maintained at 10°C inside temperature. The outside wall temperature is 35°C. The top and side walls are covered by a low conducting insulation with thermal conductivity k = 0.06 W/mK. There is no heat loss from the bottom. If heat loss through the top and side walls is to be restricted to 500W, what is the minimum thickness of insulation required? [**Ans:**108 mm]

7. A square silicon chip is of width 5mm on a side and of thickness 1mm. The chip is mounted in a substrate such that there is no heat loss from its side and back surfaces. The top surface is exposed to a coolant. The thermal conductivity of the chip is 200W/mK. If 5W are being dissipated by the chip, what is the temperature difference between its back and front surfaces? [**Ans:** 1°C]

8. The wall of a house 7m wide and 6m high is made from 0.3m thick brick with k = 0.6 W/mk. The surface temperature on the inside of the wall is 16°C and that on the outside is 6°C. Find the heat flux through the wall and the total heat loss through it. [**Ans:** -840W]

9. Consider a 3m high, 5m high and 0.3m thick wall whose thermal conductivity is 0.9W/mk. On a certain day, the temperatures of the inner and the outer surfaces of the wall are measured to be 16°C and 2°C respectively. Determine the rate of heat loss through the wall on that day. [**Ans:** 630W]

Conduction through multi-layered slabs in series

Frequently in heat conduction, heat passes through several consecutive layers of different materials. For example, in a cold store wall, heat might pass through brick, plaster, wood and cork as shown in Fig. 3.3.

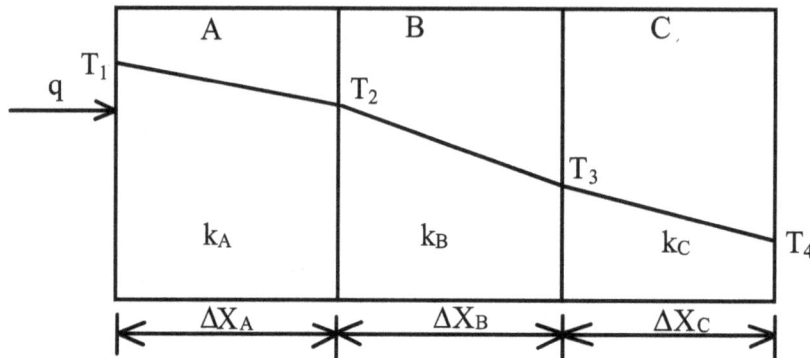

Fig. 3.3: Conduction through a multi layered slab in series

In the steady state, the same quantity of heat per unit time must pass through each layer.

$$q = \frac{A_1 \Delta T_1 k_1}{x_1} = \frac{A_2 \Delta T_2 k_2}{x_2} = \frac{A_3 \Delta T_3 k_3}{x_3} = \ldots$$

If the areas are the same,

$$A_1 = A_2 = A_3 = \ldots = A$$

Using equation 3.2;

<u>For layer A</u>

$$q = K_A \frac{A[T_1 - T_2]}{\Delta x_A}$$

Or, $T_1 - T_2 = q \cdot \dfrac{\Delta x_A}{K_A \cdot A}$ ………. (3.3)

<u>For Layer B</u>

$$q = K_B \cdot \frac{A[T_2 - T_3]}{\Delta x_B}$$

Or, $T_2 - T_3 = q \cdot \dfrac{\Delta x_B}{K_B \cdot A}$ ………. (3.4)

<u>For Layer C</u>

$$q = K_C \cdot \frac{A[T_3 - T_4]}{\Delta x_C}$$

Or, $T_3 - T_4 = q \cdot \dfrac{\Delta x_C}{K_C \cdot A}$ .... (3.5)

Adding the above equations;

$T_1-T_2 + T_2-T_3 + T_3-T_4 = q \left(\dfrac{\Delta x_A}{K_A \cdot A} + \dfrac{\Delta x_B}{K_B \cdot A} + \dfrac{\Delta x_C}{K_C \cdot A} \right)$

Or, $q = \dfrac{T_1 - T_4}{\left(\dfrac{\Delta x_A}{K_A \cdot A} + \dfrac{\Delta x_B}{K_B \cdot A} + \dfrac{\Delta x_C}{K_C \cdot A} \right)}$ .... (3.6)

Or, $q = \dfrac{(T_1 - T_4)}{(R_A + R_B + R_C)} = \dfrac{\text{Driving force}}{\text{Resistance}}$ .... (3.7)

Example 3.2: Heat transfer through a multilayered surface

1. A cold storage room is constructed of an inner layer of 12.7mm of pine, a middle layer of 101.6mm of corkboard, and an outer layer of 76.2mm of concrete. The wall surface temperature is 255.4K inside the cold room and 297.1K at the outside surface of the concrete. Use conductivities of 0.151 for pine, 0.0433 for corkboard and 0.762 W/mK for concrete. Calculate the heat loss in W for 1m² and the temperature at the interface between the wood and the corkboard.

Sol^n:

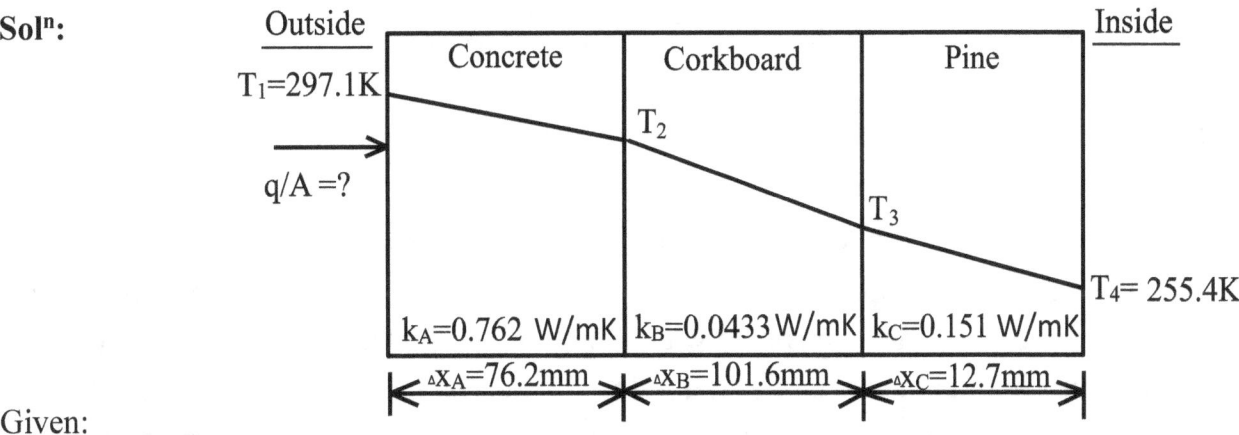

Given:

Thickness of concrete (Δx_A) = 76.2 mm = 0.0762m

Thickness of Corkboard (Δx_B) = 101.6 mm = 0.1016m

Thickness of Pine (Δx_C) = 12.7mm = 0.0127m

Outside temperature of concrete (T_1) = 297.1K

Inside temperature of room (T_4) = 255.4K

Thermal conductivity for concrete (K_A) = 0.762 W/mK

Thermal conductivity for Corkboard $(K_B) = 0.0433$ W/mK
Thermal conductivity for Pine $(K_C) = 0.151$ W/mK
Wall surface Area $(A) = 1 m^2$
Rate of heat loss $(q) = ?$
Interface temperature between Pinewood and Corkboard $(T_3) = ?$
Using equation 3.2, we can calculate resistance for each layer as:

Resistance in concrete layer $(R_A) = \dfrac{\Delta x_A}{K_A A} = 0.1$ K/W

Resistance in Corkboard layer $(R_B) = \dfrac{\Delta x_B}{K_B A} = 2.346$ K/W

Resistance in Pine layer $(R_C) = \dfrac{\Delta x_C}{K_C A} = 0.0841$ K/W

Therefore using equation 3.7, we have:

$$q = \frac{(T_1 - T_4)}{(R_A + R_B + R_C)} = \frac{(297.1 - 255.4)}{(0.1 + 2.346 + 0.0841)} = 16.48 \text{ W.}$$

Hence, the rate of heat loss per unit area is 16.48 W.

Now, using equation 3.5, we have:

$T_3 - T_4 = q \cdot \dfrac{\Delta x_C}{K_C \cdot A}$

Or, $T_3 = q \cdot \dfrac{\Delta x_C}{K_C \cdot A} + T_4 = 16.48 \times \dfrac{0.0127}{0.151 \times 1} + 255.4 = 256.79$ K.

Hence, the interface temperature between Pinewood and Corkboard is 256.79K.

Exercises on heat transfer through a multilayered slabs

1. A cold store has a wall comprising 11 cm of brick on the outside, then 7.5 cm of concrete and then 10 cm of cork. The mean temperature within the store is maintained at -18°C and the mean temperature of the outside surface of the wall is 18°C. Calculate the rate of heat transfer through the wall. The appropriate thermal conductivities are for brick, concrete and cork, respectively 0.69, 0.76 and 0.043 J m^{-1} s^{-1} °C^{-1}.

Determine also the temperature at the interfaces between the concrete and cork layers, and the brick and concrete layers. [**Ans:** q = 13.7 J s^{-1}, air/brick 18°C, brick/concrete 16°C, concrete/cork 14°C, cork/air -18°C]

2. A composite plane wall consists of two layers A and B. The thermal conductivity of layers A and B are 0.02 W/m°C and 15 W/m°C respectively. If 100 W/m² are transferred through the wall at steady state, calculate the temperature gradient in the two layers.

3. The wall of a refrigerator of 4 m² surface area consists of two metal sheets with insulation in between. The temperature of the inner wall surface is 5°C and that of the outer surface is 20°C. The thermal conductivity of the metal wall is 16 W/m°C and that of the insulation is 0.017 W/m°C. If the thickness of each metal sheet is 2 mm, calculate the thickness of the insulation that is required so that the heat transferred to the refrigerator through the wall is 10 W/m².

4. The wall of an oven consists of two metal sheets with insulation in between. The temperature of the inner wall surface is 200°C and that of the outer surface is 50°C. The thickness of each metal sheet is 2 mm, the thickness of the insulation is 5 cm, and the thermal conductivity is 16 W/m°C and 0.055 W/m°C respectively. Calculate the total resistance of the wall to heat transfer and the heat transfer losses through the wall per m² of wall area. [**Ans:** 0.90935 °C/W, 165 W]

5. Consider a two layer composite wall of copper and teflon. The copper has a thickness of 10 cm but the thickness of the teflon is to be determined. The temperature on the left boundary is equal to 200°C and on the right boundary 25°C. Determine the thickness of the teflon layer so that the heat flux is equal to 200 W/m². The thermal conductivity values for copper and teflon are 398 W/m°C and 0.25 W/m°C respectively.[**Ans:** 0.22m]

6. Determine the overall conduction heat transfer rate per unit area occurring across a furnace wall made of fire clay. Furnace wall has a thickness of 12" or a foot. The wall is insulated from outside. Thermal conductivity values for the wall and insulation materials are 0.1 W/mK and 0.01 W/mK, respectively. The furnace operates at 650°C. Average ambient temperature outside the furnace wall is 30°C and allowable temperature on the outer side of insulation is 80°C. If the air side heat transfer coefficient is 0.4 W/m²K, calculate the minimum insulation thickness requirement. [**Ans:** 0.25 m]

7. Calculate the rate of heat loss through the vertical walls of a boiler furnace of size 4 m by 3 m by 3 m high. The walls are constructed from an inner fire brick wall 25 cm thick of thermal conductivity 0.4 W/mK, a layer of ceramic blanket insulation of thermal conductivity 0.2 W/mK and 8 cm thick, and a steel protective layer of thermal conductivity 55 W/mK and 2 mm thick. The

inside temperature of the fire brick layer was measured at 600° C and the temperature of the outside of the insulation 60°C. Also find the interface temperature of layers. [**Ans:** 6320.96 W, 0.0196 K]

8. A long pipe of 0.6 m outside diameter is buried in earth with axis at a depth of 1.8 m. the surface temperature of pipe and earth are 950°C and 250°C respectively. Calculate the heat loss from the pipe per unit length. The conductivity of earth is 0.51 W/mK. [**Ans:** 90.25w/m]

9. The temperature at the inner and outer surfaces of a boiler wall made of 20 mm thick steel and covered with an insulating material of 5 mm thickness are 3000°C and 500°C respectively. If the thermal conductivities of steel and insulating material are 58W/m°C and 0.116 W/m°C respectively, determine the rate of flow through the boiler wall. [**Ans:** 5767.8 W]

10. A mild steel tank of wall thickness 10 mm contains water at 90° C. The thermal conductivity of mild steel is 50 W/m°C, and the heat transfer coefficient for inside and outside of the tank area are 2800 and 11 W/m² °C, respectively. If the atmospheric temperature is 20°C, calculate (i) The rate of heat loss per m² of the tank surface area. (ii) The temperature of the outside surface tank. [**Ans:** 765.29W/m² and 89.57°C]

11. A wall furnace is made up of inside layer of silica brick 120 mm thick covered with a layer of magnesite brick 240 mm thick. The temperatures at the inside surface of silica brick wall and outside the surface of magnesite brick wall are 725°C and 110°C respectively. The contact thermal resistance between the two walls at the interface is 0.0035°C/w per unit wall area. If thermal conductivities of silica and magnesite bricks are 1.7 W/m°C and 5.8 W/m°C, calculate the rate of heat loss per unit area of walls. [**Ans:** 5324.67 W/m]

12. A furnace walls made up of three layers, one of fire brick, one of insulating brick and one of red brick. The inner and outer surfaces are at 870° C and 40° C respectively. The respective co-efficient of thermal conduciveness of the layer are 1.0, 0.12 and 0.75 W/mK and thicknesses are 22 cm, 7.5, and 11 cm. assuming close bonding of the layer at their interfaces, find the rate of heat loss per sq.meter per hour and the interface temperatures. [**Ans:** 836.95 W/m²]

Heat conductance in parallel

Heat conductance in parallel have a sandwich construction at right angles to the direction of the heat transfer, but with heat conductances in parallel, the material surfaces are parallel to the direction of heat transfer and to each other. The heat is therefore passing through each material at the same time, instead of through one material and then the next. This is illustrated in Fig. 3.4.

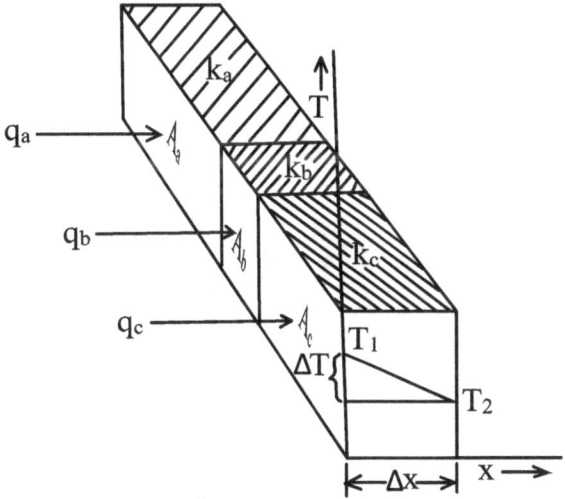

Fig. 3.4: Heat conductance in parallel

$q_a = \dfrac{A_a \Delta T K_a}{x}$; $q_b = \dfrac{A_b \Delta T K_b}{x}$; $q_c = \dfrac{A_c \Delta T K_c}{x}$

Hence, overall heat transfer rate 'q' = $q_a + q_b + q_c$ .... **(3.8)**

An example is the insulated wall of a refrigerator or an oven, in which the walls are held together by bolts. The bolts are in parallel with the direction of the heat transfer through the wall: they carry most of the heat transferred and thus account for most of the losses.

Conduction through a single hollow cylinder

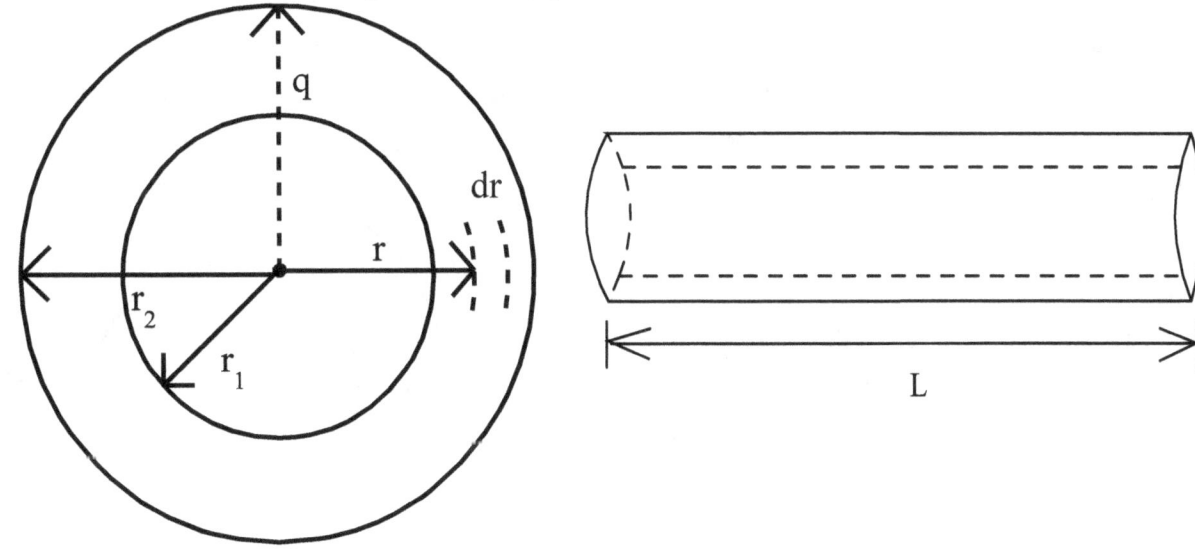

Fig. 3.5: Conduction through a hollow cylinder

Fig. 3.5 shows a hollow cylinder of inside radius r_1, outside radius r_2, length L, and thermal conductivity K. The inside and outside surfaces are maintained at constant temperatures T_1 and T_2 respectively with $T_1 > T_2$.

Rewriting Fourier's law from equation 3.1;

$$\frac{q}{A} = -K \cdot \frac{dT}{dr}$$

Since $A = 2\pi rL$, and;

Integrating the above equation for dT from T_1 to T_2 where dr goes from r_1 to r_2:

$$q = \frac{K \cdot 2\pi L (T_1 - T_2)}{\ln\left(\frac{r_2}{r_1}\right)} \quad \ldots\ldots (3.9)$$

Multiplying numerator and denominator by $(r_2 - r_1)$, we get;

$$q = \frac{K \cdot A_{lm} \cdot (T_1 - T_2)}{(r_2 - r_1)} \quad \text{Where,} \quad A_{lm} = \frac{2\pi L (r_2 - r_1)}{\ln\left(\frac{r_2}{r_1}\right)} = \text{Log mean area} \ldots\ldots (3.10)$$

Or, $$q = \frac{(T_1 - T_2)}{\left\{\frac{(r_2 - r_1)}{K \cdot A_{lm}}\right\}} = \frac{(T_1 - T_2)}{R} = \frac{\text{Driving force}}{\text{Resistance}} \quad \ldots\ldots (3.11)$$

Example 3.3: Heat transfer through a stainless steel pipe

1. Hot water is transferred through a stainless steel pipe of 0.04 m inside diameter and 5 m length. The inside wall temperature is 90°C, the outside surface temperature is 88°C, the thermal conductivity of stainless steel is 16 W/m°C, and the wall thickness is 2 mm. Calculate the heat losses if the system is at steady state.

Sol^n:

Given,

Internal radius of pipe $(r_1) = 0.04/2 = 0.02$ m

Thickness of pipe $(t) = 2$ mm $= 0.002$ m

∴ Outer radius of pipe $(r_2) = r_1 + t = 0.022$ m.

Length of the pipe $(L) = 5$ m

Inside wall temperature of pipe $(T_1) = 90$°C

Outside surface temperature $(T_2) = 88$°C

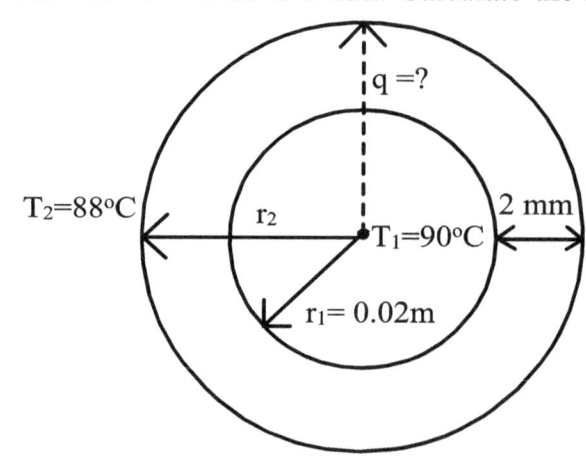

Thermal conductivity of steel (K) = 16 W/m°C

Rate of heat loss (q) =?

From equation 3.9, we can write:

$$q = \frac{K \cdot 2\pi L (T_1 - T_2)}{\ln\left(\frac{r_2}{r_1}\right)} = \frac{16 \times 2 \times \pi \times 5 \,(90-88)}{\ln\left(\frac{0.022}{0.02}\right)} = 10526 \text{ W}.$$

Hence, the heat losses to the surrounding is 10526 W.

Exercises on heat transfer through a single layered cylinder

1. A thick walled cylindrical tubing of hard rubber having an inside radius of 10mm and an outside radius of 25mm is being used as a cooling coil in a bath. Ice water is flowing rapidly inside and inside wall temperature is 274.9 K. The outside surface temperature is 297.1 K. A total of 14.65W must be removed from the bath by the cooling coil. How many meter of tubing is needed? The thermal conductivity (K) is 0.15W/mK.

2. A spherical shaped vessel of 1.2 m outer diameter is 100 mm thick. Find the rate of heat leakage, if the temperature difference between the inner and outer surfaces is 200°C. Thermal conductivity of material is 0.3 kJ /mh°C. [**Ans:** 574.4 W]

Conduction through multi-layered cylinders in series

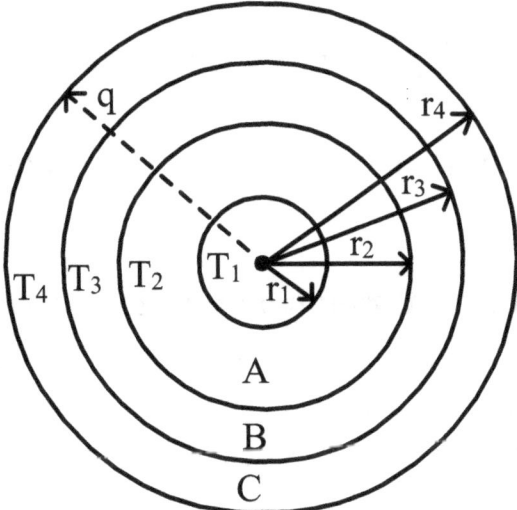

Fig. 3.6: Conduction through multi-layered cylinders in series

Consider a multi-layered cylinder with 3 layers A, B and C with temperatures of T_1, T_2 and T_3 respectively and T_4 on the outer surface of C. The radius of 3 layers are r_2, r_3 and r_4 respectively with r_1 as the inner radius. The heat 'q' is flowing from inside of the tube to the outside surface as shown in Fig. 3.6. Solving for heat transfer in each layer separately; and

From Fourier's law using equation 3.11;

For layer A

$$q = \frac{(T_1 - T_2)}{\left\{\frac{(r_2 - r_1)}{K_A \cdot A_{Alm}}\right\}} \quad \ldots\ldots\ldots (3.12)$$

For layer B

$$q = \frac{(T_2 - T_3)}{\left\{\frac{(r_3 - r_2)}{K_B \cdot A_{Blm}}\right\}} \quad \ldots\ldots\ldots (3.13)$$

For layer C

$$q = \frac{(T_3 - T_4)}{\left\{\frac{(r_4 - r_3)}{K_C \cdot A_{clm}}\right\}} \quad \ldots\ldots\ldots (3.14)$$

Solving for 'q' from each layer;

$$q = \frac{T_1 - T_4}{(R_A + R_B + R_C)} = \frac{\text{Driving force}}{\text{Resistance}} \quad \ldots\ldots\ldots (3.15)$$

Where, $R_A = \dfrac{(r_2 - r_1)}{K_A \cdot A_{Alm}}$; and so on for R_B and R_C .... (3.16)

Example 3.4: Heat transfer through a multi-layered steel pipe

1. A thick walled tube of stainless steel having K= 21.63 W/mK with dimensions of 0.0254m ID and 0.0508m OD is covered with a 0.0254m thick layer of an insulation having K= 0.2423 W/mK. The inside wall temperature of pipe is 811 K and the outside surface of the insulation is at 310.8 K. For a 0.305m length of pipe, calculate the heat loss and also the temperature at the interface between the metal and the insulation.

Sol^n:

Given,

Thermal conductivity of steel (K_A) = 21.63 W/mK

Inside radius of pipe (r_1) = 0.0254/2 = 0.0127m

Outside radius of pipe (r_2) = 0.0508/2 = 0.0254m

Thickness of insulation over pipe (t) = 0.0254m

∴ Outer radius of insulation (r_3) = r_2 + t = 0.0508m

Thermal conductivity of insulation (K_B) = 0.2423 W/mK

Inside wall temperature of pipe (T_1) = 811K

Outside surface temperature of insulation (T_3) = 310.8K

Length of pipe (L) = 0.305m

Heat loss to the outside (q) =?

Interfacial temperature between metal and insulation (T_2) =?

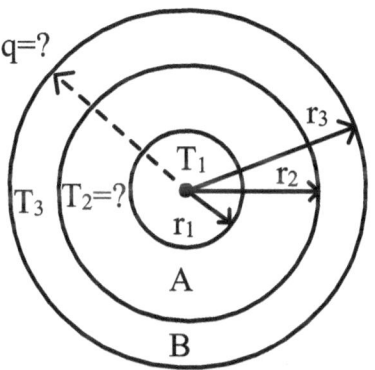

Using equation 3.10, calculate the log mean area as:

$$A_{Alm} = \frac{2\pi L(r_2 - r_1)}{\ln\left(\frac{r_2}{r_1}\right)} = \frac{811 - T_2}{0.01673} = 0.0351 \text{ m}^2$$

$$A_{Blm} = \frac{2\pi L(r_3 - r_2)}{\ln\left(\frac{r_3}{r_2}\right)} = \frac{2 \times \pi \times 5 \times (0.0508 - 0.0254)}{\ln\left(\frac{0.0508}{0.0254}\right)} = 0.0703 \text{ m}^2$$

Using equation 3.11, calculate the resistance for both layers as:

$$R_A = \frac{(r_2 - r_1)}{K_A \cdot A_{Alm}} = \frac{(0.0254 - 0.0127)}{21.63 \times 0.0351} = 0.01673 \text{ K/W}$$

$$R_B = \frac{(r_3 - r_2)}{K_B \cdot A_{Blm}} = \frac{(0.0508 - 0.0254)}{0.2423 \times 0.0703} = 1.491 \text{ K/W}$$

Therefore, the heat transfer rate is given from equation 3.15;

$$q = \frac{T_1 - T_3}{(R_A + R_B)} = \frac{811 - 310.8}{(0.01673 + 1.491)} = 331.7 \text{ W}$$

Similarly using equation 3.12, we can calculate interfacial temperature T_2 as:

$$q = \frac{T_1 - T_2}{R_A}$$

Or, $331.7 = \dfrac{811 - T_2}{0.01673}$

∴ $T_2 = 805.5$ K.

Hence, the heat loss is 331.7 W and the temperature at the interface between the metal and the insulation is 805.5 K.

Exercises on heat transfer through a multilayered cylinders

1. If an insulation of 2 cm thickness with thermal conductivity equal to 0.02 W/m°C is wrapped around the pipe of previous example so that the outside surface temperature of the insulation is 35°C, while the inside wall temperature is still 90°C, what would be the heat loss? What will be the outside surface metal wall temperature T_2?

2. An spherical container of negligible thickness holding a hot fluid at 1400 and having an outer diameter of 0.4 m is insulated with three layers of each 50 mm thick insulation of $k_1 = 0.02$: $k_2 = 0.06$ and $k_3 = 0.16$ W/mK. (Starting from inside). The outside surface temperature is 300°C. Determine (i) the heat loss, and (ii) Interface temperatures of insulating layers. [**Ans:** 21.57 W and 35.09°C].

3. A steel tube with 5 cm ID, 7.6 cm OD and k=15W/m°C is covered with an insulative covering of thickness 2 cm and k 0.2W/m°C . A hot gas at 330°C with h = 400W/m²°C flows inside the tube. The outer surface of the insulation is exposed to cooler air at 30°C with h = 60 W/m²°C. Calculate the heat loss from the tube to the air for 10 m of the tube and the temperature drops resulting from the thermal resistances of the hot gas flow, the steel tube, the insulation layer and the outside air. [**Ans:** 7451.72 W, 11.859K, 3.310K, 34.07K].

4. A steel pipe (K=45.0 W/m.K) having a 0.05m O.D is covered with a 0.042 m thick layer of magnesia (K=0.07W/m.K) which in turn covered with a 0.024 m layer of fiberglass insulation (K=0.048 W/m.K). The pipe wall outside temperature is 370K and the outer surface temperature of the fiberglass is 305K. What is the interfacial temperature between the magnesia and fiberglass? Also calculate the steady state heat transfer. [**Ans:** 19.959 W/m and 325.26K].

5. A 15 cm outer diameter steam pipe is covered with 5 cm high temperature insulation (k = 0.85 W/m°C) and 4 cm of low temperature (k = 0.72 W/m°C). The steam is at 500°C and ambient air is at 40°C. Neglecting thermal resistance of steam and air sides and metal wall calculate the heat

loss from 100m length of the pipe. Also find temperature drop across the insulation. [**Ans:** 2929.75W/m]

6. Water flows through a cast steel pipe (k=50 W/mK) with an outer diameter of 104 mm and 2 mm wall thickness. Calculate the heat loss by convection and conduction per meter length of uninsulated pipe when the water temperature is 15°C, the outside air temperature is -10°C, the water side heat transfer coefficient is 30kW/m^{2}K and the outside heat transfer coefficient is 20 W/m^{2}K. Calculate the corresponding heat loss when the pipe is lagged with insulation having an outer diameter of 300mm and thermal conductivity of K = 0.05 W/mK. [**Ans:** 163.3 W/m and 7.3 W/m]

7. Water at 80°C is pumped through 100m of stainless steel pipe, K = 16 W/mK of inner and outer radii 47mm and 50mm respectively. The heat transfer coefficient due to water is 2000W/m^{2}K. the outer surface of the pipe loses heat by convection to air at 20°C and the heat transfer coefficient is 200W/m^{2}K. Calculate the heat flow through the pipe. Also calculate the heat flow through the pipe when a layer of insulation, K= 0.1 W/mK and 50mm radial thickness is wrapped around the pipe. [**Ans:** 0.329×10^6 W and 5.39×10^3 W]

8. A steam pipe 8m long has an external diameter of 100mm and it is covered by lagging 50mm thick. The pipe contains steam at 198°C and the temperature of the atmosphere surrounding the pipe is 18°C. The thermal conductivity of the lagging is 0.15 W/mK. The surface heat transfer coefficient is 10 W/m^{2}K. Assuming the pipe has the same uniform temperature as the steam, calculate the heat loss and the surface temperature of the lagging. [**Ans:** 6.1W and 18.04°C]

9. A spherical tank 2m diameter contains liquefied fuel gas. It is covered in insulation 120mm thick with a thermal conductivity of 0.025 W/mK. The surface heat transfer coefficient between the lagging and the surrounding air is 30 W/m^{2}K. the air is at 25°C and the liquid is -125°C. Assume the inner surface temperature is the same as the liquid. Calculate the heat transfer rate required to keep the liquid at a constant temperature and surface temperature of the insulation. [**Ans:** 437W and 24.1°C]

Unsteady state heat transfer

In food process engineering, heat transfer is very often in the unsteady state, in which temperatures are changing and materials are warming or cooling. Unfortunately, study of heat flow under these conditions is complicated. In fact, it is the subject for study in a substantial branch of applied mathematics, involving finding solutions for the Fourier equation written in terms of partial differentials in three dimensions. There are some cases that can be simplified and handled by elementary methods, and also charts have been prepared which can be used to obtain numerical solutions under some conditions of practical importance.

A simple case of unsteady state heat transfer arises from the heating or cooling of solid bodies made from good thermal conductors, for example a long cylinder, such as a meat sausage or a metal bar, being cooled in air.

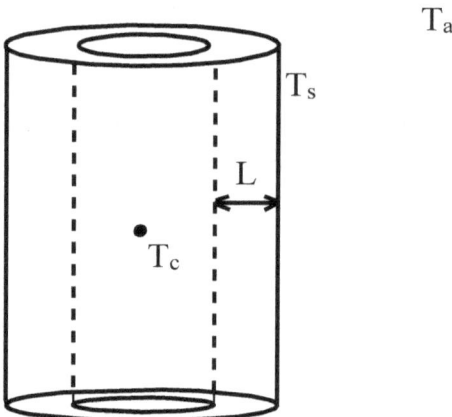

Fig. 3.7: Hot cylinder being cooled in air

Consider a long hot cylinder being cooled in air as given in figure 3.7 above. The rate at which heat is transferred to the air from the surface of the cylinder is given by;

$$q = \frac{dQ}{dt} = h_s A (T_s - T_a) \quad \text{.......... (3.17)}$$

where, T_a = air temperature, T_s = surface temperature and h_s = heat transfer coefficient at the surface. The heat is being lost to the surface from the interior of the cylinder by conduction. Hence, applying Fourier's law:

$$\frac{dQ}{dt} = \frac{K A (T_c - T_s)}{L} \quad \text{.......... (3.18)}$$

where, L is the thickness of the cylinder.

From equation 3.17 and 3.18;

$$h_s A (T_s - T_a) = \frac{K A (T_c - T_s)}{L} \quad \text{.......... (3.19)}$$

$$\frac{h_s L}{K} = \frac{(T_c - T_s)}{(T_s - T_a)} \quad \text{.......... (3.20)}$$

Biot number (Bi)

The term $\frac{h_s L}{K}$ is called the biot number in equation 3.20.

$$\text{Biot number} = \frac{\text{Heat transfer coefficient at the surface}}{\text{Heat conductance to the centre of the solid}}$$

It is the dimensionless number and simply the ratio of heat transfer resistances inside of and at the surface of the body. In general, problems involving small biot numbers (Much smaller than 1) re thermally simple due to uniform temperature fields inside the body. Biot numbers much larger than 1 produce more difficult problems due to non- uniformity of temperature fields within the object.

Biot number smaller than 0.2 labels a substance as thermally thin and temperature can be assumed to be constant throughout the materials volume.

Lumped heat capacity method (K>>>h)

From equation 3.17, we can write;

$dQ = h_s A [T_s - T_a] dt$

This loss of heat cools the cylinder in accordance with the usual specific heat equation;

$dQ = m C_p dT = C_p \rho V dT$ .... (3.21)

Where, C_p = Specific heat of the material of the cylinder, ρ & V are the density and volume of the cylinder. Since the heat passing through the surface must equal the heat loss from the cylinder, hence from equation 3.17 & 3.21;

$C_p \rho V dT = h_s A [T_s - T_a] dt$

Or, $\dfrac{dT}{(T_s - T_a)} = \dfrac{h_s A \, dt}{C_p \rho V}$ .... (3.22)

Integrating the above equation 3.22 for T_s from T_1 to T_2 i.e the initial and final temperature of the cylinder during the cooling period from 0 to t;

$$\frac{[T_2 - T_a]}{[T_1 - T_a]} = e^{\left(\frac{-h_s A t}{C_p \rho V}\right)} \quad \text{.......... (3.23)}$$

The term $C_p \rho V$ is called the lumped heat capacitance of the system.

Fourier number (Fo)

It is a dimensionless number that characterizes heat conduction. It is the ratio of heat conduction rate to the rate of thermal energy storage. Together with Biot number, it characterizes transient conduction problems.

$$F_o = \frac{\alpha t}{L^2} \quad \cdots\cdots\cdots (3.24)$$

Where, α = thermal diffusivity in m^2/s.
t = characteristic time in seconds
L = Length through which conduction results.

Heat Transfer by convection

Convection heat transfer is the transfer of energy by the mass movement of groups of molecules. It is restricted to liquids and gases, as mass molecular movement does not occur at an appreciable speed in solids. It cannot be mathematically predicted as easily as can transfer by conduction or radiation and so its study is largely based on experimental results rather than on theory. The most satisfactory convection heat transfer formulae are relationships between dimensionless groups of physical quantities. Furthermore, since the laws of molecular transport govern both heat flow and viscosity, convection heat transfer and fluid friction are closely related to each other.

Convection coefficients will be studied under two sections, firstly, natural convection in which movements occur due to density differences on heating or cooling; and secondly, forced convection, in which an external source of energy is applied to create movement. In many practical cases, both mechanisms occur together.

Natural convection

Heat transfer by natural convection occurs when a fluid is in contact with a surface hotter or colder than itself. As the fluid is heated or cooled it changes its density. This difference in density causes movement in the fluid that has been heated or cooled and causes the heat transfer to continue.

There are many examples of natural convection in the food industry. Convection is significant when hot surfaces, such as retorts which may be vertical or horizontal cylinders, are exposed with or without insulation to colder ambient air. It occurs when food is placed inside a chiller or freezer store in which circulation is not assisted by fans. Convection is important when material is placed in ovens without fans and afterwards when the cooked material is removed to cool in air.

It has been found that natural convection rates depend upon the physical constants of the fluid, density ρ, viscosity μ, thermal conductivity k, specific heat at constant pressure Cp and coefficient of thermal expansion β (beta) which for gases = $1/T$ by Charles' Law. Other factors that also affect convection-heat transfer are, some linear dimension of the system, diameter D or length L, a temperature difference term, ΔT, and the gravitational acceleration g since it is density differences acted upon by gravity that create circulation. Heat transfer rates are expressed in terms of a convection heat transfer coefficient h_c, which is part of the general surface coefficient h_s.

Experimentally, if has been shown that convection heat transfer can be described in terms of these factors grouped in dimensionless numbers which are known by the names of eminent workers in this field:

Nusselt number (Nu) = $\dfrac{h_c L}{k}$ .... (3.25)

Prandtl number (Pr) = $\dfrac{c_p \mu}{k}$ .... (3.26)

Grashof number (Gr) = $\dfrac{L^3 \rho^2 g \beta \Delta T}{\mu^2}$ .... (3.27)

and in some cases a length ratio (L/D). L is length of height which equals Area/ Perimeter.

If we assume that these ratios can be related by a simple power function we can then write the most general equation for natural convection:

$$(Nu) = K(Pr)^k (Gr)^m (L/D)^n \quad .... (3.28)$$

Experimental work has evaluated K, k, m, n, under various conditions. Once K, k, m, n, are known for a particular case, together with the appropriate physical characteristics of the fluid, the Nusselt number can be calculated. From the Nusselt number we can find h_c and so determine the rate of convection-heat transfer by applying equation 3.29. In natural convection equations, the values of the physical constants of the fluid are taken at the mean temperature between the surface and the bulk fluid. The Nusselt and Biot numbers look similar: they differ in that for Nusselt, k and h both refer to the fluid, for Biot k is in the solid and h is in the fluid.

Exercises on dimensionless numbers in convection

1. Calculate the Prandtl number for the following:
a. Water at 20°C : μ=1.002 ×10⁻³ kg/ms, Cp=4.183kJ/kgK and k=0.603W/mK [**Ans:**6.95]
b. Mercury at 20°C: μ= 1520×10⁻⁶ kg/ms, Cp= 0.139 kJ/kgK and k= 0.0081kW/mK [**Ans:**0.0261]
c. Liquid sodium at 400K: μ= 420×10⁻⁶ kg/ms, Cp= 1369 J/kgK and k= 86W/mK [**Ans:**0.0067]

d. Engine oil at 60°C: μ= 8.36×10⁻² kg/ms, Cp= 2035 J/kgK and k= 0.141 W/mK [**Ans:**1207]

2. Calculate the appropriate Grashof numbers for the following:

a. A central heating radiator, 0.6m high with a surface temperature of 75°C in a room at 18°C (ρ= 1.2 kg/m³, Pr= 0.72 and μ= 1.8×10⁻⁵ kg/ms) [**Ans:** 1.84×10⁹]

b. A horizontal oil sump with a surface temperature of 40°C, 0.4m long and 0.2m wide containing oil at 75°C (ρ = 854 kg/m³, Pr= 546, β=0.7×10⁻³ K⁻¹ and μ=3.56×10⁻²kg/ms) [**Ans:** 4.1×10⁴]

c. The external surface of a heating coil, 30 mm diameter, having a surface temperature of 80°C in water at 20°C (=1000kg/m³, Pr= 6.95, = 0.227×10⁻³ K⁻¹ and μ= 1×10⁻³ kg/ms) [**Ans:** 3.6×10⁶]

3. Calculate the Nusselt number when there is a flow of gas (Pr=0.71, μ= 4.63×10⁻⁵ kg/ms and C_P= 1175 J/kgK) over a turbine blade of chord length 20 mm where the average heat transfer coefficient is 1000 W/m²K. [**Ans:** 261]

Forced convection

When a fluid is forced past a solid body and heat is transferred between the fluid and the body, this is called forced convection heat transfer. Examples in the food industry are in the forced-convection ovens for baking bread, in blast and fluidized freezing, in ice-cream hardening rooms, in agitated retorts, in meat chillers. In all of these, foodstuffs of various geometrical shapes are heated or cooled by a surrounding fluid, which is moved relative to them by external means.

The fluid is constantly being replaced, and the rates of heat transfer are, therefore, higher than for natural convection. Also, as might be expected, the higher the velocity of the fluid the higher the rate of heat transfer. In the case of low velocities, where rates of natural convection heat transfer are comparable to those of forced convection heat transfer, the Grashof number is still significant. But in general the influence of natural circulation, depending as it does on coefficients of thermal expansion and on the gravitational acceleration, is replaced by dependence on circulation velocities and the Reynold's number.

As with natural convection, the results are substantially based on experiment and are grouped to deal with various commonly met situations such as fluids flowing in pipes, outside pipes, etc.

Surface heat transfer coefficients

Newton found, experimentally, that the rate of cooling of the surface of a solid, immersed in a colder fluid, was proportional to the difference between the temperature of the surface of the solid and the temperature of the cooling fluid. This is known as Newton's Law of Cooling, and it can be expressed by the equation:

$q = h_s A(T_a - T_s)$ .... **(3.29)**

where h_s is called the surface heat transfer coefficient, T_a is the temperature of the cooling fluid and T_s is the temperature at the surface of the solid. The surface heat transfer coefficient can be regarded as the conductance of a hypothetical surface film of the cooling medium of thickness x_f such that

$h_s = \dfrac{k_f}{x_f}$ where, k_f is the thermal conductivity of the cooling medium.

Exercises on heat transfer by convection

1. Air flows over a rectangular plate having dimensions 0.5 m x 0.25 m. The free stream temperature of the air is 300°C. At steady state, the plate temperature is 40°C. If the convective heat transfer coefficient is 250 W/m²K, determine the heat transfer rate from the air to one side of the plate. [**Ans**: 8125 W]

2. Calculate the heat transfer per square meter between a fluid with a bulk temperature of 66°C with a wall with surface temperature of 25°C given h= 5W/m²K. [**Ans**: 205W]

Combined conduction and convection (In a plane slab)

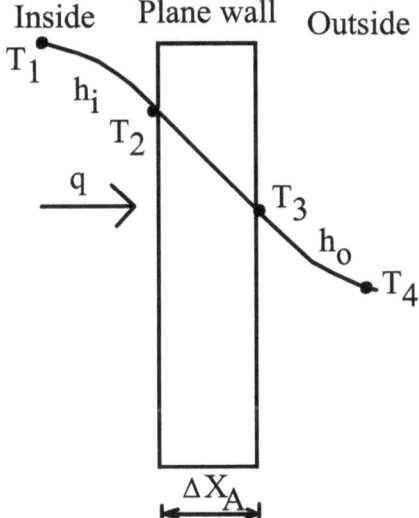

Fig. 3.8: Combined conduction and convection in a plane slab

Consider a plane wall with hot fluid at temperature T_1 on the inside surface and cold fluid at T_4 on the outside surface. The convective coefficient on the outside is h_o and h_i on the inside (W/m²K) as shown in Fig. 3.8.

If q amount of heat is being transferred from hot liquid to cold liquid through plane wall of thickness Δx_A, then the rate can be calculated as;

$$q = h_i A (T_1 - T_2) = \frac{K_A \cdot A (T_2 - T_3)}{\Delta x_A} = h_o A (T_3 - T_4) \quad \ldots\ldots\ldots (3.30)$$

Solving for ΔT:

$$T_1 - T_2 = \frac{q}{h_i A} \quad \ldots\ldots\ldots (3.31)$$

$$T_2 - T_3 = \frac{q \, \Delta x_A}{K_A A} \quad \ldots\ldots\ldots (3.32)$$

$$T_3 - T_4 = \frac{q}{h_o A} \quad \ldots\ldots\ldots (3.33)$$

Adding all the equations and solving for q:

$$q = \frac{T_1 - T_4}{\left(\dfrac{1}{h_i A} + \dfrac{\Delta x_A}{K_A A} + \dfrac{1}{h_o A}\right)} = \frac{(T_1 - T_4)}{(R_1 + R_2 + R_3)} \quad \ldots\ldots\ldots (3.34)$$

The overall heat transfer by combined conduction and convection is often expressed in terms of an overall heat transfer coefficient "U". It dictates the amount of heat that passes through 1 m² of the partition per 1 °C of differential temperature.

$q = U A \Delta T_{overall}$, where ΔT overall $= T_1 - T_4$

Therefore,

$$U = \frac{1}{\left(\dfrac{1}{h_i} + \dfrac{\Delta x_A}{K_A} + \dfrac{1}{h_o}\right)} \text{ in W/m}^2\text{K} \ldots\ldots\ldots(3.35)$$

Similarly, the overall heat transfer rate through cylinder is;

$$q = \frac{(T_1 - T_4)}{\left(\dfrac{1}{h_i A_i} + \dfrac{(r_o - r_i)}{K_A \cdot A_{lm}} + \dfrac{1}{h_o A_o}\right)} \quad \ldots\ldots\ldots (3.36)$$

where, r_o = Outside radius and r_i = inside radius.

Example 3.5: Overall heat transfer coefficient in slab

1. Calculate the respective U values for a wall made from either (a) 10 cm of brick of thermal conductivity 0.7 J m^{-1}s^{-1} °C^{-1}, or (b) 1.3mm of aluminium sheet, conductivity 208 J m^{-1} s^{-1} °C^{-1}. Surface heat-transfer coefficients are on the one side 9.8 and on the other 40 J m^{-2} s^{-1} °C^{-1}.

Sol^n:
Given,
Surface heat transfer on the inside (h_i) = 9.8 Jm^{-2}s^{-1}m^{-2}
Surface heat transfer on the outside (h_o) = 40 Jm^{-2}s^{-1}m^{-2}
Thermal conductivity of brick (K_A) = 0.7 J m^{-1}s^{-1} °C^{-1}
Thermal conductivity of aluminium (K_B) = 208 Jm^{-1}s^{-1} °C^{-1}
Thickness of brick (Δx_A) = 10 cm = 0.1m
Thickness of aluminium sheet (Δx_B) = 1.3 mm = 0.0013m
Overall heat transfer coefficient (U) = ?

From equation 3.35, we can calculate respective 'U' values as:

a. For Brick layer:

$$U = \frac{1}{\left(\frac{1}{h_i} + \frac{\Delta x_A}{K_A} + \frac{1}{h_o}\right)} = \frac{1}{\left(\frac{1}{9.8} + \frac{0.1}{0.7} + \frac{1}{40}\right)} = 3.7 \text{ Jm}^{-2}\text{s}^{-1}\text{°C}^{-1}$$

b. For Aluminium layer:

$$U = \frac{1}{\left(\frac{1}{h_i} + \frac{\Delta x_B}{K_B} + \frac{1}{h_o}\right)} = \frac{1}{\left(\frac{1}{9.8} + \frac{0.0013}{208} + \frac{1}{40}\right)} = 7.7 \text{ Jm}^{-2}\text{s}^{-1}\text{°C}^{-1}$$

Hence, the overall heat transfer coefficient for brick and aluminium sheet are found to be 3.7 and 7.7 Jm^{-2}s^{-1}°C^{-1}.

Exercises on overall heat transfer coefficient

1. Two jacketed kettles made of carbon steel (λ = 50 W/m°C [28.9 Btu/hr-ft°F] with an inner wall thickness of 15mm [0.049 ft] are used to heat water. One uses hot water as the heat source, while the other uses steam. Assuming heat transfer coefficients of 1000 W/m^2°C [176 Btu/hr-ft^2°F] for the water being heated, 3000 W/m^2°C [528 Btu/hr-ft^2°F] for hot water, and 10000 W/m^2°C [1761 Btu/hr-ft^2°F] for steam, calculate the U values for both heating processes. [**Ans:** 714 W/m^2°C and 398 W/m^2°C].

2. Calculate the heat transfer per square meter between a fluid with a bulk temperature of 160°C and another with a bulk temperature of 15°C with a wall between them made of two layers A and B both 50mm thick. The surface heat transfer coefficient for the hot fluid is 5 W/m²K and for the cold fluid 3 W/m²K. The thermal conductivity of layer A is 20 W/mK and for B it is 0.5 W/mK. Also calculate the overall heat transfer coefficient. [**Ans:** 228.2 W and 1.573 W/m²K]

3. Calculate the heat transfer through a steel plate with water on one side at 90°C and air on the other at 15°C. The plate has an area of 1.5 m² and it is 20mm thick. The thermal conductivity is 60 W/mK. The surface heat transfer coefficients for the air and water respectively are 0.006 W/m²K and 0.08 W/m²K. Calculate the heat loss and the overall heat transfer coefficient. [**Ans:** 0.628 W and 5.581 × 10⁻³ W/m²K]

4. A wall with an area of 25 m² is made up of four layers. On the inside is plaster 15mm thick, then there is brick 100mm thick, then insulation 60mm thick and finally brick 100mm thick. The thermal conductivity of plaster is 0.1 W/mK. The thermal conductivity of brick is 0.6 W/mK. The thermal conductivity of the insulation is 0.08 W/mK. The one side of the wall is in contact with air at 22°C and the other with air at -5°C. The surface heat transfer coefficient for both surfaces is 0.006 W/m²K. Calculate the heat loss and the overall heat transfer coefficient. [**Ans:** 1.79W and 3 × 10⁻³ W/m²K]

5. A steel tube of 1 mm wall thickness is being used to condense ammonia, using cooling water outside the pipe in a refrigeration plant. If the water side heat transfer coefficient is estimated at 1750 J m⁻² s⁻¹ °C⁻¹ and the thermal conductivity of steel is 45 J m⁻¹ s⁻¹ °C⁻¹, calculate the overall heat-transfer coefficient assuming the ammonia condensing coefficient, 6000 J m⁻² s⁻¹ °C⁻¹. [**Ans:** 1300 J m⁻² s⁻¹ °C⁻¹].

6. The wall of a building is a multi-layered composite consisting of brick (100-mm layer), a 100-mm layer of glass fiber (paper faced 28kg/m²), a 10-mm layer of gypsum plaster (vermiculite), and a 6-mm layer of pine panel. If h_{inside} is 10W/m²K and $h_{outside}$ is 70W/m²K, calculate the total thermal resistance and the overall coefficient for heat transfer.[**Ans:** 2.93 m²K/W and 0.341 W/m²K]

Fouling factors

In actual practice, heat transfer surfaces do not remain clean. Dirt, soot, scales and other deposits form on one or both sides of the tubes of an exchanger and on other heat transfer surfaces. These deposits offer additional resistance to the flow of heat and reduces the overall heat transfer coefficient 'U'.

The effect of such deposits and fouling is usually taken care of in design by adding a term for the resistance of the fouling on the inside and outside of the tube as:

$$U_i = \frac{1}{\left\{\dfrac{1}{h_i} + \dfrac{1}{h_{di}} + \dfrac{(r_o - r_i)A_i}{K_A \cdot A_{Alm}} + \dfrac{A_i}{A_o h_o} + \dfrac{A_i}{A_o h_{do}}\right\}} \quad \ldots\ldots (3.37)$$

Where h_{di} is the fouling coefficient for the inside and h_{do}, the fouling coefficient for the outside of the tube in W/ m²K. A similar expression can be written for U_o using the same values.

Table 3.1: Typical fouling coefficient in W/ m²K

Fouling factors	Fouling coefficients
Distilled and sea water	11350
City water	5680
Muddy water	1990- 2840
Gases	2840
Vapourizing liquids	2840
Vegetables and gas oils	1990

Heat transfer to boiling liquids

Boiling is associated with transformation of liquid to vapour at a solid/liquid interface due to convection heat transfer from the solid.

Agitation of fluid by vapour bubbles provides for large convection coefficients and hence large heat fluxes at low-to moderate surface-to-fluid temperature differences.

Special form of Newton's law of cooling

$q_s = h(T_s - T_{sat}) = h\, \Delta T_e$ where, T_{sat} is the saturation temperature of the liquid, and $\Delta T_e = (T_s - T_{sat})$ is the excess temperature.

The boiling curve

The boiling curve reveals range of conditions associated with saturated pool boiling on a q_s vs. ΔT_e plot as given in 3.9.

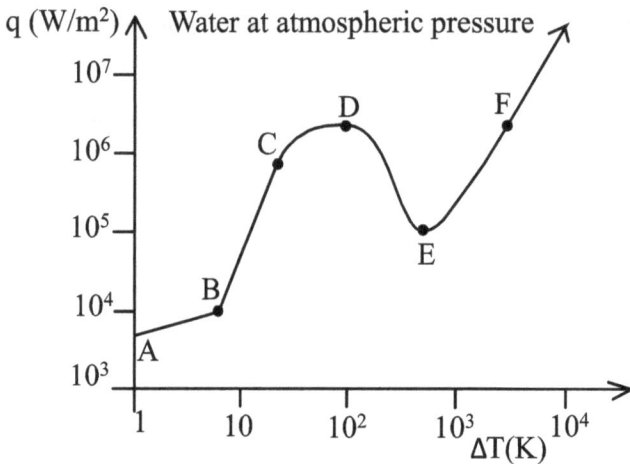

Fig. 3.9: Boiling curve of water at atmospheric pressure

A-B : Pure convection with liquid rising to surface for evaporation.

B-C : Nucleate boiling with bubbles condensing in liquid.

C-D : Nucleate boiling with bubbles rising to surface.

D : Peak temperature.

D-E : Partial nucleate boiling and unstable film boiling.

E : Film boiling is stabilized.

E-F : Radiation becomes a dominant mechanism for heat transfer.

In the first region of the plot (A-B), at low temperature drops, the mechanism of boiling is essentially that of heat transfer to a liquid in natural convection. The variation of h with $\Delta T^{0.5}$ is approximately the same as that for natural convection to horizontal plate or cylinders. Very few bubbles formed are released from the surface of metal and rise without appreciably disturbing the natural convection.

In the region (B-D) of nucleate boiling, boiling for a ΔT of about (5-25) K, the rate of bubble production increases so that the velocity of circulation of liquid increases. The heat transfer coefficient h increases rapidly and is proportional to ΔT^2 to ΔT^3 in this region.

In the region (D-E) of transition boiling, many bubbles are formed so quickly that they tend to coalescence and form a layer of insulating vapour. Increasing the ΔT increases the thickness of layer and the heat flux and h drops as ΔT increases.

In the region (E-F) of film boiling, bubbles detach themselves regularly and rises upward. At higher ΔT values, the radiation through the vapour layer next to the surface helps to increase q/A and h. Similar shaped graph will be obtained for all types of geometry of heating surfaces.

Heat transfer from condensing vapours

The rate of heat transfer obtained when a vapour is condensing to a liquid is very often important. In particular, it occurs in the food industry in steam-heated vessels where the steam condenses and gives up its heat; and in distillation and evaporation where the vapours produced must be condensed. In condensation, the latent heat of vapourization is given up at constant temperature, the boiling temperature of the liquid.

Two generalized equations have been obtained:

For condensation on vertical tubes or plane surfaces

$$h_v = 0.94\left\{\left(\frac{k^3\rho^2 g}{\mu}\right) \times \left(\frac{\lambda}{L\Delta T}\right)\right\}^{0.25} \qquad \text{.......... (3.38)}$$

where λ (lambda) is the latent heat of the condensing liquid in J kg^{-1}, L is the height of the plate or tube and the other symbols have their usual meanings.

For condensation on a horizontal tube

$$h_h = 0.72\left\{\left(\frac{k^3\rho^2 g}{\mu}\right) \times \left(\frac{\lambda}{D\Delta T}\right)\right\}^{0.25} \qquad \text{.......... (3.39)}$$

where D is the diameter of the tube.

These equations apply to condensation in which the condensed liquid forms a film on the condenser surface. This is called film condensation: it is the most usual form and is assumed to occur in the absence of evidence to the contrary. However, in some cases the condensation occurs in drops that remain on the surface and then fall off without spreading a condensate film over the whole surface. Since the condensate film itself offers heat transfer resistance, film condensation heat transfer rates would be expected to be lower than drop condensation heat transfer rates and this has been found to be true. Surface heat-transfer rates for drop condensation may be as much as ten times as high as the rates for film condensation.

The contamination of the condensing vapour by other vapours, which do not condense under the condenser conditions, can have a profound effect on overall coefficients. Examples of a non-condensing vapour are air in the vapours from an evaporator and in the jacket of a steam pan. The adverse effect of non-condensable vapours on overall heat transfer coefficients is due to the difference between the normal range of condensing heat transfer coefficients, 1200-12,000 $J\ m^{-2}\ s^{-1}\ °C^{-1}$, and the normal range of gas heat transfer coefficients with natural convection or low velocities, of about 6 $J\ m^{-2}\ s^{-1}\ °C^{-1}$.

Uncertainties make calculation of condensation coefficients difficult, and for many purposes it is near enough to assume the following coefficients:

For condensing steam   12,000 $J\ m^{-2}\ s^{-1}\ °C^{-1}$
For condensing ammonia   6,000 $J\ m^{-2}\ s^{-1}\ °C^{-1}$
For condensing organic liquids 1,200 $J\ m^{-2}\ s^{-1}\ °C^{-1}$

The heat-transfer coefficient for steam with 3% air falls to about 3500 $J\ m^{-2}\ s^{-1}\ °C^{-1}$, and with 6% air to about 1200 $J\ m^{-2}\ s^{-1}\ °C^{-1}$.

Heat exchangers

In a heat exchanger, heat energy is transferred from one body or fluid stream to another. In the design of heat exchange equipment, heat transfer equations are applied to calculate this transfer of energy so as to carry it out efficiently and under controlled conditions. The equipment goes under many names, such as boilers, pasteurizers, jacketed pans, freezers, air heaters, cookers, ovens and so on. The range is too great to list completely. Heat exchangers are found widely scattered throughout the food process industry.

Continuous-flow heat exchangers

It is very often convenient to use heat exchangers in which one or both of the materials that are exchanging heat are fluids, flowing continuously through the equipment and acquiring or giving up heat in passing as shown in Fig. 3.10 , 3.11 and 3.12.

One of the fluids is usually passed through pipes or tubes, and the other fluid stream is passed round or across these. At any point in the equipment, the local temperature differences and the heat transfer coefficients control the rate of heat exchange.

Countercurrent flow heat exchangers

The temperature difference between the two liquids is best utilized if they flow in opposite directions through the heat exchanger (Fig.3.10). The cold product then meets the cold heating medium at the inlet, and a progressively warmer medium as it passes through the heat exchanger. During the passage, the product is gradually heated so that the temperature is always only a few degrees below that of the heating medium at the corresponding point. This type of arrangement is called countercurrent flow.

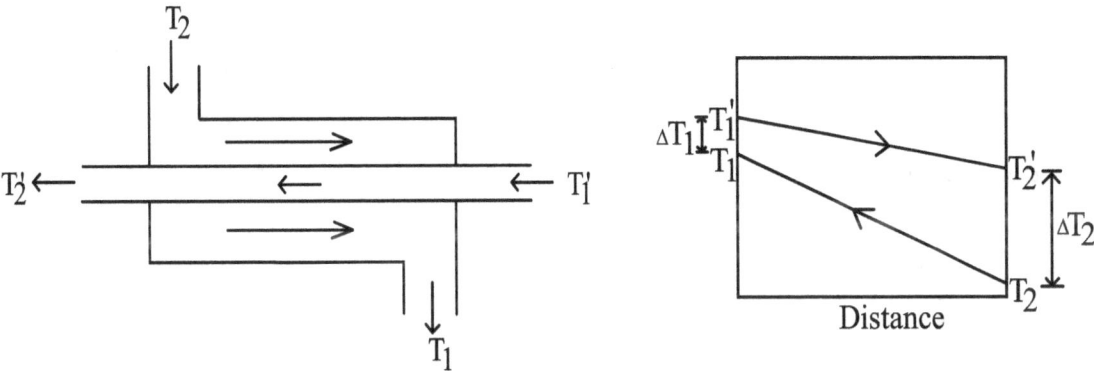

Fig. 3.10: Temperature profiles for heat transfer in a heat exchanger with countercurrent flow

Concurrent flow heat exchangers

With the opposite arrangement, concurrent flow (Fig.3.11), both liquids enter the heat exchanger from the same end and flow in the same direction. In concurrent flow, it is impossible to heat the product to a temperature higher than that which would be obtained if the product and the heating medium were mixed. This limitation does not apply in countercurrent flow; the product can be heated to within two or three degrees of the inlet temperature of the heating medium.

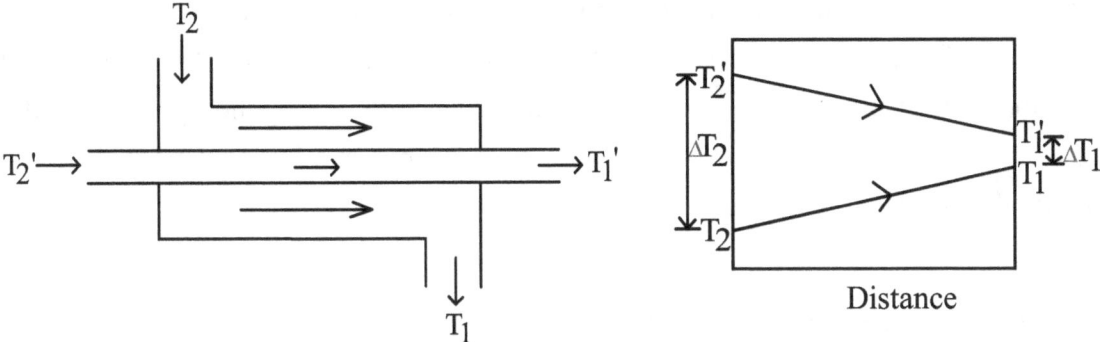

Fig. 3.11: Temperature profiles for heat transfer in a heat exchanger with concurrent flow

Cross flow heat exchangers

The fluids flow at right angles to each other, called cross flow.

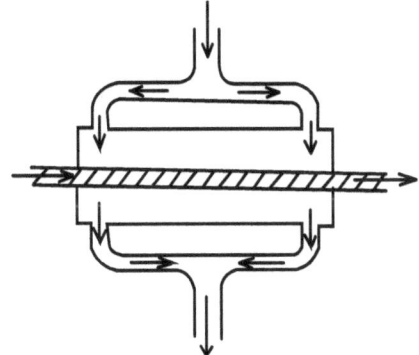

Fig. 3.12: Cross flow heat exchanger

Log mean temperature difference [LMTD]

It has already been mentioned that there must be a difference in temperature between the two media for heat transfer to take place. The differential temperature is the driving force. The greater the difference in temperature, the more heat is transferred and the smaller the heat exchanger needed. For sensitive products there are, however, limits to how great a difference can be used.

The differential temperature can vary through the heat exchanger. A mean value, LTMD, is used for calculation. It is denoted by ΔT_{lm} and can be calculated by the following formula, using the denominations in Fig. 3.10.

$$\Delta T_{lm} = \frac{(T_2' - T_2) - (T_1' - T_1)}{\ln\left(\frac{T_2' - T_2}{T_1' - T_1}\right)} = \frac{\Delta T_2 - \Delta T_1}{\ln\left(\frac{\Delta T_2}{\Delta T_1}\right)} \quad \ldots\ldots\ldots (3.40)$$

NTU method

The Number of Transfer Units (NTU) Method is used to calculate the rate of heat transfer in heat exchangers (especially counter current exchangers) when there is insufficient information to calculate the Log-Mean Temperature Difference (LMTD). In heat exchanger analysis, if the fluid inlet and outlet temperatures are specified or can be determined by simple energy balance, the LMTD method can be used; but when these temperatures are not available, the NTU or the Effectiveness method is used.

For any heat exchanger it can be shown that its effectiveness:

$$E = f\left(NTU, \frac{C_{min}}{C_{max}}\right) \ldots\ldots\ldots (3.41)$$

C_{min} and C_{max} are the heat capacity rates of cold and hot fluids respectively.
and the number of transfer units, NTU:

$$NTU = \frac{UA}{C_{min}} \ldots\ldots\ldots (3.42)$$

where U is the overall heat transfer coefficient and A is the heat transfer area.

Shell and tube heat exchangers

If larger heat transfers are required, a shell and tube heat exchanger is used which is the most important type in use in the process industries. In these type, the flow are continuous. Many tubes in parallel are used where one fluid flows inside these tubes.

The tubes arranged in a bundle are enclosed in a single shell and the other fluid flows outside the tube on the shell side.

1 Shell and 1 tube pass heat exchanger

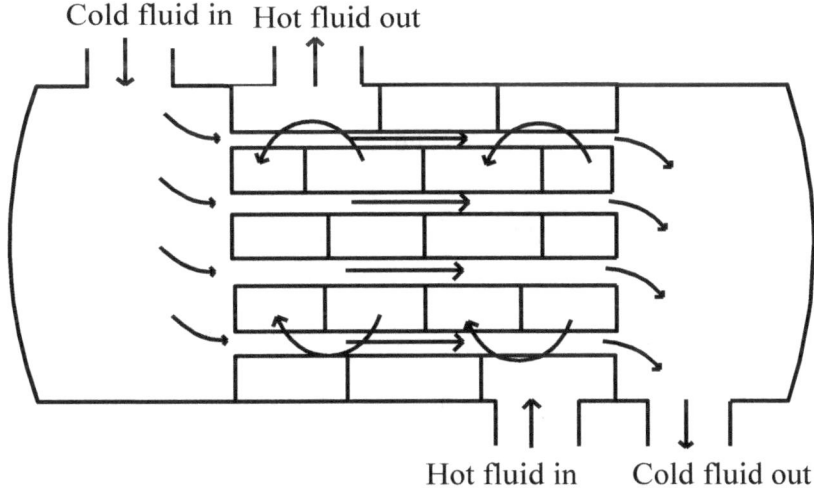

Fig. 3.13: 1 Shell and 1 tube pass heat exchanger

As shown in Fig. 3.13, the cold fluid enters and flows inside all the tubes in parallel in one pass. The hot fluid enters at the other end and flows counter flow across the outside of the tubes. Cross baffles are used so that the fluid is forced to flow perpendicular across the tube tank rather than parallel with it. The added turbulence generated by this cross flow increases the shell side heat transfer coefficient.

1 Shell pass and 2 tube pass heat exchanger

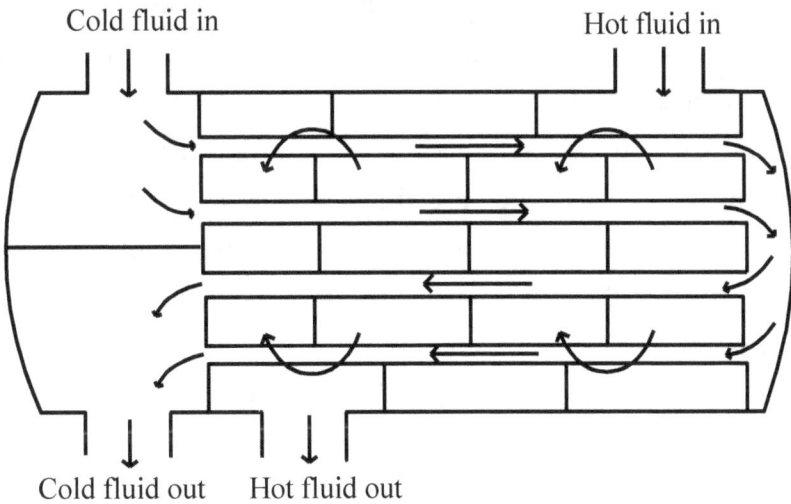

Fig. 3.14: 1 Shell and 2 tube pass heat exchanger

As shown in Fig. 3.14, the liquid on the tube side flows in 2 passes as shown and the shell side liquid flows in one pass. In the first pass of the tube side, the cold fluid is flowing counter flow to the hot shell side fluid. In the second pass of the tube side, the cold fluid flows is parallel with the hot fluid.

Plate heat exchangers

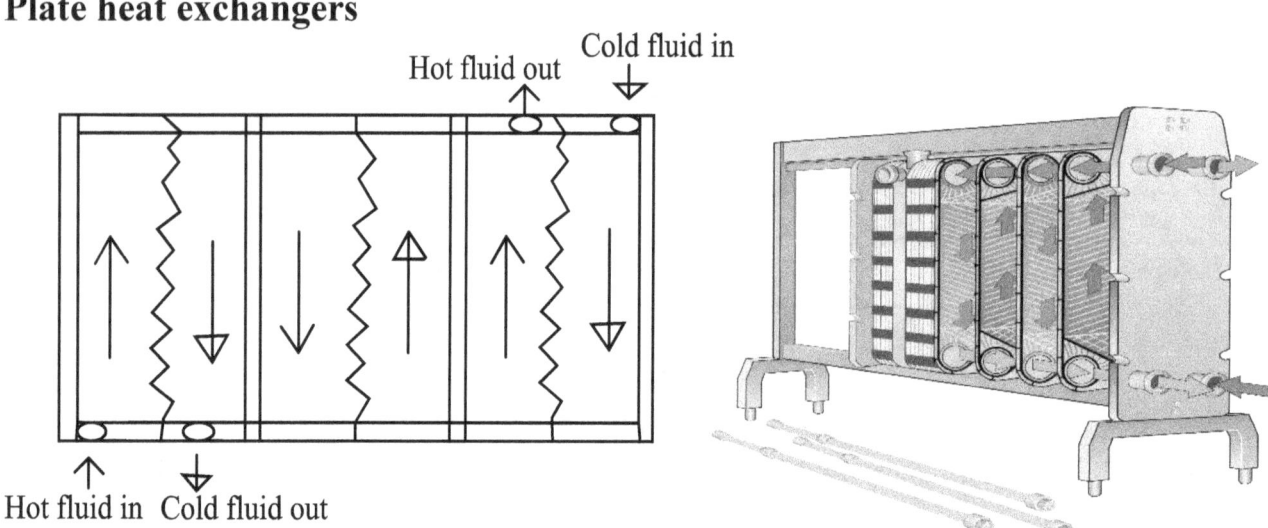

Fig. 3.15: Principles of flow and heat transfer in a plate heat exchanger

For applications which runs at moderate temperature and pressure, an alternative to the shell- tube exchanger is the gasketed plate heat exchanger which consists of many corrugated stainless steel sheets separated by polymer gaskets and clamped in a steel frame. Inlet portals and slots in the gaskets direct the hot and cold fluid to the alternate spaces between the plates.

As shown in Fig. 3.15, the plate heat exchanger consists of a pack of corrugated metal plates with portholes for the passage of the two fluids between which heat transfer will take place. The plate pack is assembled between a fixed frame plate and a movable pressure plate and compressed by tightening bolts. The plates are fitted with a gasket which seals the interplate channel and directs the fluids into alternate channels. The number of plates is determined by the flow rate, physical properties of the fluids, pressure drop and temperature program. The plate corrugations promote fluid turbulence and support the plates against differential pressure. The plate and the pressure plate are suspended from an upper carrying bar and located by a lower guiding bar, both of which are fixed to a support column. Connections are located in the frame plate or, if either or both fluids make more than a single pass within the unit, in the frame and pressure plates.

Channels are formed between the plates and the corner ports are arranged so that the two media flow through alternate channels. The heat is transferred through the plate between the channels, and complete counter-current flow is created for highest possible efficiency. The corrugation of the plates provides the passage between the plates, supports each plate against the adjacent one and enhances the turbulence, resulting in efficient heat transfer.

The corrugation induce turbulence for improved heat transfer, and each plate is supported by multiple contact with adjoining plates, which have a different pattern or angle of corrugation. The space between plates is equal to the depth of the corrugation and is usually 2-5 mm. Overall heat transfer coefficients are of the order of 2400-6000 J m^{-2} s^{-1} °C^{-1}. It is used mainly for pasteurization and chilling of raw milk.

Chapter 4: Evaporation

Introduction
The process of evaporation is employed in the food industry primarily as a means of bulk and weight reduction in fluids. It is used extensively in the dairy industry to concentrate milk, in the fruit juice industry to produce fruit juice concentrates, in the manufacture of jams, jellies and preserves, and to raise the solid content necessary for gelling; and in the sugar industry to concentrate the solids of dilute solutions prior to spray or freeze drying. The process of evaporation involves the application of heat to vaporize water at the boiling point. Separation is achieved by exploiting the difference in volatility between water and solutes. The concentration of a solution by boiling- off solvent has three major applications in the food industry:

a. pre- concentration of a liquid prior to further processing, e.g. before spray drying, drum drying, crystallization, etc.

b. Reduction of liquid volume to reduce storage, packaging and transport costs.

c. To increase the concentration of soluble solids in food materials as an aid to preservation, e.g. in sweetened condensed milk manufacture.

Factors affecting evaporation

Concentration in the liquid
Usually, the liquid feed to an evaporator is relatively dilute, so its viscosity is low, similar to water, and relatively high heat transfer coefficients are obtained. As evaporation proceeds, the solutions become very concentrated and quite viscous, causing the heat transfer coefficient to drop markedly. Adequate circulation and/or turbulence must be present to keep the coefficient from becoming too low.

Solubility
As solution are heated and concentration of the solute or salt increases, the solubility limit of the material in solution may be exceeded and crystals may form. This may limit the maximum concentration in solution which can be obtained by evaporation.

Temperature sensitivity of material
Many produces especially food and other biological materials, may be temperature sensitive and degrade at higher temperatures or after prolonged heating. Such products are pharmaceutical products; food products such as milk, orange juice, and vegetable extracts; and fine orange juice.

Foaming or frothing

In some cases, materials composed of caustic solutions, food solutions such as skim milk, and some fatty acid solutions form a foam or froth during boiling. This foam accompanies the vapour coming out of the evaporator and entrainment losses occur.

Pressure and temperature

The boiling point of the solution is related to the pressure of the system. The higher the pressure of the evaporator, the higher the temperature at boiling. Also, as the concentration of the dissolved material in solution increases by evaporation, the temperature of boiling may rise. This phenomenon is called boiling point rise or elevation.

Scale deposition and materials of construction

Some solutions deposit solid materials called scale on the heating surfaces. These could be formed by decomposition products or solubility decreases. The result is that the overall heat transfer coefficient decreases and the evaporator must eventually be cleaned.

Boiling-point elevation

As evaporation proceeds, the liquor remaining in the evaporator becomes more concentrated and its boiling point will rise. The extent of the boiling-point elevation depends upon the nature of the material being evaporated and upon the concentration changes that are produced. The extent of the rise can be predicted by Raoult's Law, which leads to:

$\Delta T = kx$ .... **(4.1)**

Where, ΔT is the boiling point elevation, x is the mole fraction of the solute and k is a constant of proportionality.

In multiple effect evaporators, where the effects are fed in series, the boiling points will rise from one effect to the next as the concentrations rise. Relatively less of the apparent temperature drops are available for heat transfer, although boiling points are higher, as the condensing temperature of the vapour in the steam chest of the next effect is still that of the pure vapour. Boiling-point elevation complicates evaporator analysis but heat balances can still be drawn up along the lines indicated previously. Often foodstuffs are made up from large molecules in solution, in which boiling-point elevation can to a greater extent be ignored. As the concentrations rise, the viscosity of the liquor also rises. The increase in the viscosity of the liquor affects the heat transfer and it often imposes a limit on the extent of evaporation that is practicable. There is no straight forward method of predicting the extent of the boiling-point elevation in the concentrated solutions that are met in some evaporators in practical situations. Many solutions have their boiling points at some concentrations tabulated in the literature, and these can be extended by the use of a relationship known as Duhring's rule. Duhring's rule states that the ratio of the temperatures at which two solutions (one of which can be pure water) exert the same vapour pressure is constant.

Thus, if we take the vapour pressure/temperature relation of a reference liquid, usually water and if we know two points on the vapour pressure-temperature curve of the solution that is being evaporated, the boiling points of the solution to be evaporated at various pressures can be read off from the diagram called a Duhring plot. The Duhring plot will give the boiling point of solutions of various concentrations by interpolation, and at various pressures by proceeding along a line of constant composition. A Duhring plot of the boiling points of sodium chloride solutions is given in Fig. 4.1.

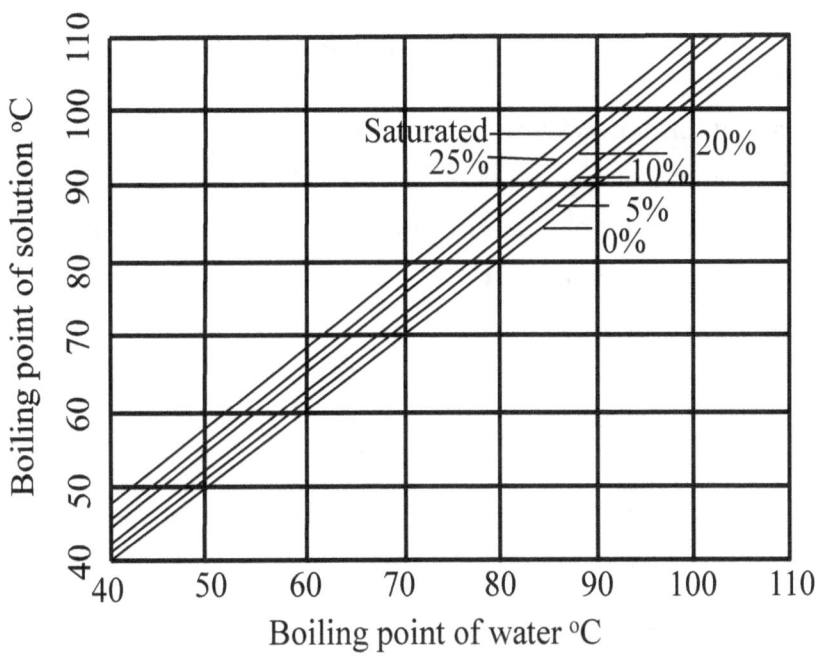

Fig. 4.1: Duhring plot for boiling point of sodium chloride solutions

By finding the boiling points of salt solutions of various concentrations under two pressures, the Duhring lines can then, also, be filled in for solutions of these concentrations. Such lines are also on Fig. 4.1. Intermediate concentrations can be estimated by interpolation and so the complete range of boiling points at any desired concentration, and under any given pressure, can be determined.

Evaporation Equipments
Open pans

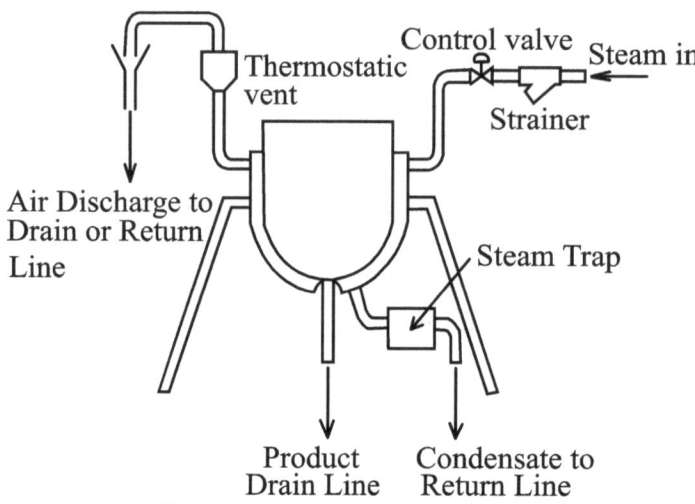

Fig. 4.2: Open pan evaporator

The most elementary form of evaporator consists of an open pan in which the liquid is boiled. Heat can be supplied through a steam jacket or through coils, and scrapers or paddles may be fitted to provide agitation. Such evaporators are simple and low in capital cost, but they are expensive in their running cost as heat economy is poor.

Horizontal-tube evaporators

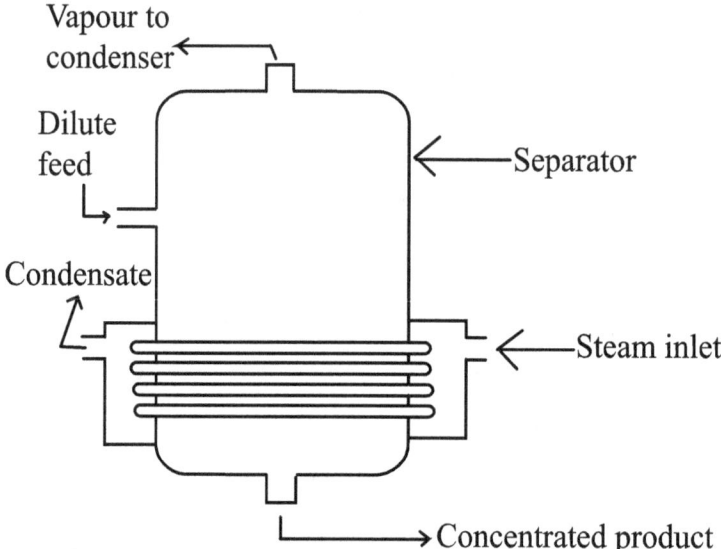

Fig. 4.3: Horizontal tube evaporator

The horizontal-tube evaporator is a development of the open pan, in which the pan is closed in, generally in a vertical cylinder. The heating tubes are arranged in a horizontal bundle immersed in the liquid at the bottom of the cylinder. Liquid circulation is rather poor in this type of evaporator.

This type of unit was originally intended for the evaporation of non-scaling, non-foaming, low viscosity liquids.

Vertical-tube evaporators

By using vertical, rather than horizontal tubes, the natural circulation of the heated liquid can be made to give good heat transfer. Recirculation of the liquid is through a large "downcomer" so that the liquors rise through the vertical tubes about 5-8 cm diameter, boil in the space just above the upper tube plate and recirculate through the downcomers. The hydrostatic head reduces boiling on the lower tubes, which are covered by the circulating liquid. The length to diameter ratio of the tubes is of the order of 15:1.

a. Basket evaporator

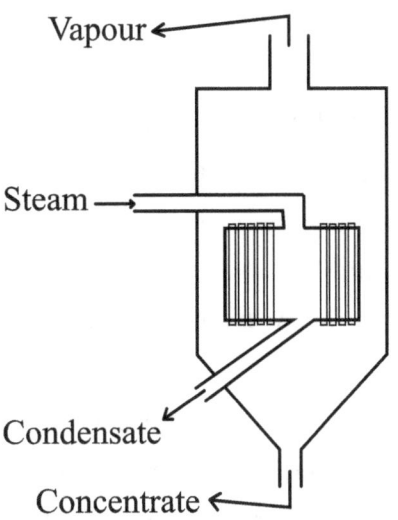

Fig. 4.4: Basket evaporator

The basket evaporator shown in Fig. 4.4 is a variant of the calandria evaporator in which the steam chest is contained in a basket suspended in the lower part of the evaporator, and recirculation occurs through the annular space round the basket.

b. Long tube evaporators

Tall slender vertical tubes may be used for evaporators as shown in the Fig. 4.5. The tubes, which may have a length to diameter ratio of the order of 100:1, pass vertically upward inside the steam chest. The liquid may either pass down through the tubes, called a falling-film evaporator, or be carried up by the evaporating liquor in which case it is called a climbing-film evaporator. Evaporation occurs on the walls of the tubes. Because circulation rates are high and the surface films are thin, good conditions are obtained for the concentration of heat sensitive liquids due to high heat transfer rates and short heating times.

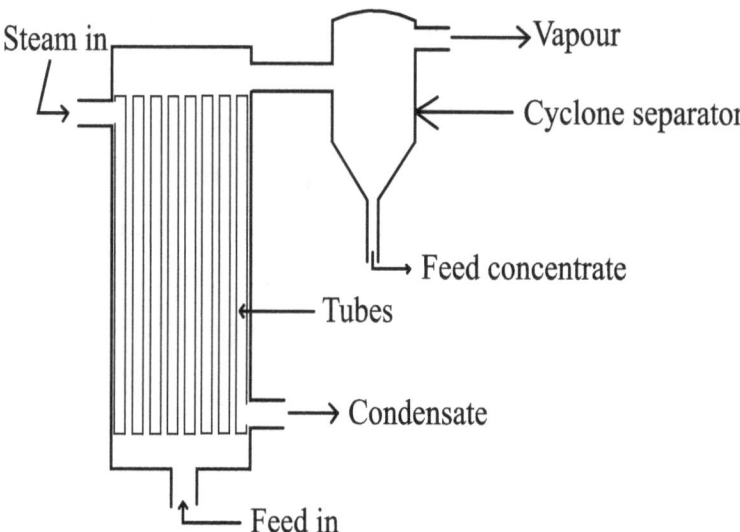

Fig. 4.5: Climbing film long tube evaporator

In the climbing-film evaporator, as the liquid boils on the inside of the tube slugs of vapour form and this vapour carries up the remaining liquid which continues to boil. Tube diameters are of the order of 2.5 to 5 cm, contact times may be as low as 5-10 sec.

Overall heat- transfer coefficients may be up to five times as great as from a heated surface immersed in a boiling liquid. In the falling-film type, the tube diameters are rather greater, about 8 cm, and these are specifically suitable for viscous liquids.

Forced circulation evaporators

The heat transfer coefficients from condensing steam are high, so that the major resistance to heat flow in an evaporator is usually in the liquid film. Tubes are generally made of metals with a high thermal conductivity, though scale formation may occur on the tubes which reduce the tube conductance.

The liquid-film coefficients can be increased by improving the circulation of the liquid and by increasing its velocity of flow across the heating surfaces. Pumps, or impellers, can be fitted in the liquid circuit to help with this. Using pump circulation, the heat exchange surface can be divorced from the boiling and separating sections of the evaporator, as shown in the Fig. 4.6. Alternatively, impeller blades may be inserted into flow passages such as the downcomer of a calandria- type evaporator.

Forced circulation is used particularly with viscous liquids: it may also be worth consideration for expensive heat exchange surfaces when these are required because of corrosion or hygiene requirements. In this case it pays to obtain the greatest possible heat flow through each square metre of heat exchange surface.

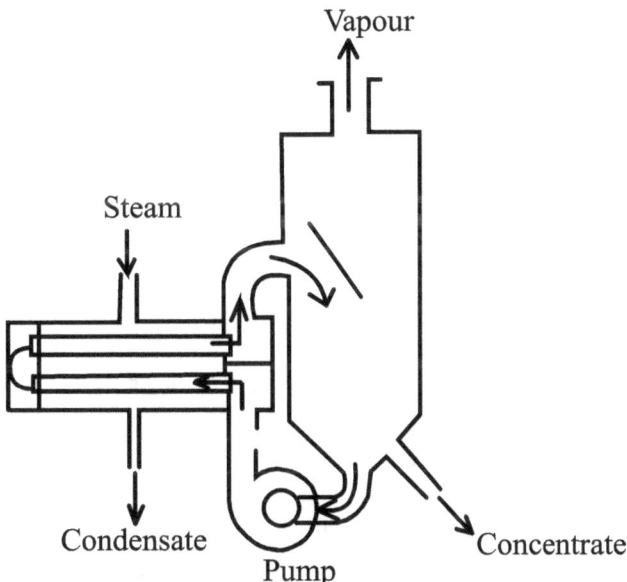

Fig. 4.6: Forced circulation evaporator

Also under the heading of forced-circulation evaporators are various scraped surface and agitated film evaporators. In one type the material to be evaporated passes down over the interior walls of a heated cylinder and it is scraped by rotating scraper blades to maintain a thin film, high heat transfer and a short and controlled residence time exposed to heat.

Plate evaporators

The plate heat exchanger can be adapted for use as an evaporator. The spacings can be increased between the plates and appropriate passages provided so that the much larger volume of the vapours, when compared with the liquid, can be accommodated. Plate evaporators can provide good heat transfer and also ease of cleaning.

Evaporation for heat-sensitive liquids

Many food products with volatile flavor constituents retain more of these if they are evaporated under conditions favoring short contact times with the hot surfaces. This can be achieved for solutions of low viscosity by climbing- and falling-film evaporators, either tubular or plate types. As the viscosity increases, for example at higher concentrations, mechanical transport across heated surfaces is used to advantage. Methods include mechanically scraped surfaces, and the flow of the solutions over heated spinning surfaces. Under such conditions residence times can be fractions of a minute and when combined with a working vacuum as low as can reasonably be maintained, volatiles retention can be maximized.

Mass and heat balance in single effect evaporator

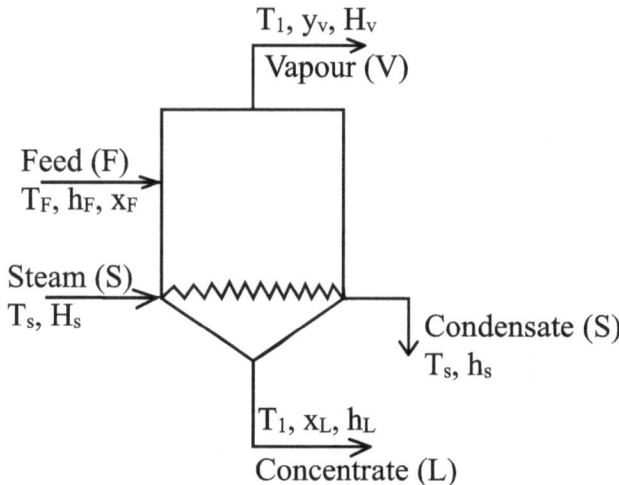

Fig. 4.7: Mass and heat balance in a single effect evaporator

The feed to the evaporator is F kg/hr having solid content of x_F mass fraction, temperature T_F and Enthalpy h_F J/kg. Coming out as a liquid is the concentrated liquid L kg/hr having solids content of x_L, temperature T_1 and Enthalpy h_L. The vapour V kg/hr is given off as pure solvent having solid content $y_v=0$, temperature T_1 and Enthalpy H_v. Saturated steam entering is S kg/hr and has a temperature of T_s and Enthalpy of H_s. The condensed steam leaving of S kg/hr is assumed usually to be at the saturation temperature T_s with enthalpy h_s. This means that steam gives off only its latent heat, λ

$$\lambda = H_s - h_s \quad \ldots \ldots \ldots (4.2)$$

Since, the vapour V is in equilibrium with the liquid L, the temperature of vapour and liquid are the same. Also the pressure P_1 is the saturated vapour pressure of the liquid of composition x_L and boiling temperature T_1. Assume no boiling point rise.

a. <u>For material balance:</u>

At the steady state

Rate of mass in = Rate of mass out

Or, $F = L + V \ldots \ldots \ldots (4.3)$

For a solute balance alone,

$F \cdot x_F = L \cdot x_L \ldots \ldots \ldots (4.4)$

b. <u>For heat balance:</u>

Total heat entering = Total heat leaving

i.e. Heat in feed + Heat in Steam = Heat in concentrated liquid + Heat in vapour + Heat in condensate

Or, $F \cdot h_F + S \cdot H_s = L \cdot h_L + V \cdot H_v + S \cdot h_s$

Or, $F \cdot h_F + S(H_s - h_s) = L \cdot h_L + V \cdot H_v$

Or, $F \cdot h_F + S \cdot \lambda = L \cdot h_L + V \cdot H_v$ .... **(4.5)**

The heat q transferred to the evaporator is then

$$q = S(H_s - h_s) = S \cdot \lambda$$

The overall rate of heat transfer, q for single effect evaporator is given by

$q = UA \, \Delta T$ .... **(4.6)**

where, U= Overall heat transfer coefficient (W/m²K)

A= Heat transfer surface area (m²)

$\Delta T = T_s - T_1$, in Kelvin.

Example 4.1: Heat transfer area of single effect evaporator

1. A single effect evaporator concentrates 9072 kg/hr of a 1 wt.% salt solution entering at 311K to a final concentration of 1.5 wt.%. The steam supplied is saturated at 383.2K and the vapour at the vapour space of evaporator is at 373.2K. Calculate the heat transfer area required if the overall heat transfer coefficient is 1704 W/m²K ; the latent heat of vapour at 373.2K and that of steam at 383.2K and the heat capacity of the feed are 2257 kJ/kg, 2230 kJ/kg and 4.14 kJ/kgK respectively. Assume that the solution has same boiling point as water.

Solⁿ:

Given,

Feed rate (F) = 9072 kg/hr

Feed concⁿ (x_F) = 0.01

Feed temperature (T_F) =311K

Steam temperature (T_S) = 383.2K

Vapour temperature (T_1) = 373.2K

Concentrate temperature (T_1) = 373.2K

Concⁿ of concentrate (x_L) = 0.015

Overall heat transfer coefficient (U) =1704W/m²K

Heat capacity of feed (C_F) = 4.14 kJ/kgK

Enthalpy of concentrate (h_L) =0 [As the solⁿ has same boiling point as water]

Latent heat of vapour (λ_v) = 2257 kJ/kg

∴ Enthalpy of vapour (H_v)= 2257 kJ/kg as, $\lambda_v = H_v - h_L$

Latent heat of steam (λ) = 2230 kJ/kg

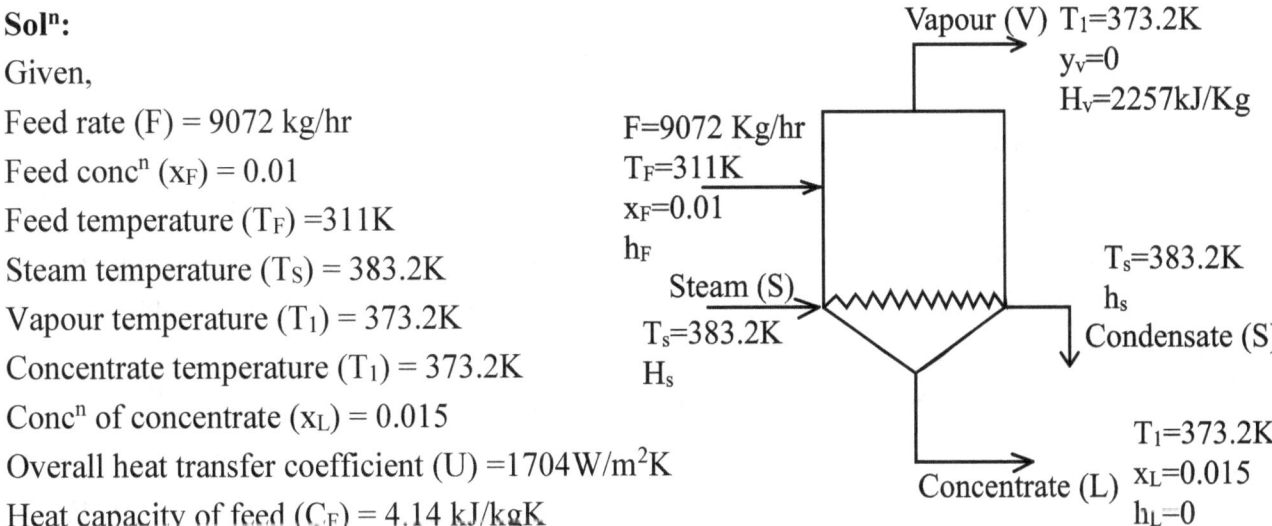

∴ Enthalpy of feed (h_F) = C_F ($T_F - T_1$) = -257.508 kg/kJ

a. Making overall mass balance over an evaporator, as of equation 4.3;

F = L + V

Or, 9072 = L + V.......... (4.7)

b. Making a component mass balance (of salt concentration), as of equation 4.4;

$F.x_F = L.x_L + V.y_v$

Or, 9072 ×0.01 = L × 0.015 + 0

∴ L = 6048 kg/hr and using equation (1), V = 3024 kg/hr.

c. Making a heat balance in an evaporator, as of equation 4.5;

$F. h_F + S. \lambda = L. h_L + V. H_v$

Or, 9072 × -257.508 + S × 2230 = 6048 ×0 + 3024 ×2257

∴ S = 4108.19 kg/hr

The rate of heat transfer to the evaporator (q) is given as:

q = S. λ

   = 2544800 W

Using equation 4.6, q can also be calculated as:

q = UAΔT = UA($T_S - T_1$)

Or, A = $\dfrac{2544800.16}{1704 \times 10}$ = 149.34

Hence, the heat transfer area required is found to be 149.34 m².

Exercises on single effect evaporator

1. A single effect evaporator is required to concentrate a solution from 10% solids to 30% solids at the rate of 250 kg of feed per hour. If the pressure in the evaporator is 77 kPa absolute, and if steam is available at 200 kPa gauge, calculate the quantity of steam required per hour and the area of heat transfer surface if the overall heat transfer coefficient is 1700 J m^{-2} s^{-1} °C^{-1}.

Assume that the temperature of the feed is 18°C and that the boiling point of the solution under the pressure of 77 kPa absolute is 91°C. Assume, also, that the specific heat of the solution is the same as for water, that is 4.186 x 10³ J kg^{-1}°C^{-1}, and the latent heat of vapourization of the solution is the same as that for water under the same conditions.

From steam tables, the condensing temperature of steam at 200 kPa (gauge)[300 kPa absolute] is 134°C and latent heat 2164 kJ kg^{-1}; the condensing temperature at 77 kPa (abs.) is 91°C and latent heat is 2281 kJ kg^{-1}. [**Ans:** 1.74 m²]

2. A single effect evaporator is to be used to concentrate a food solution containing 15% (by mass) dissolved solids to 50% solids. The feed stream enters the evaporator at 291 K with a feed rate of 1.0 kg s^{-1}. Steam is available at a pressure of 2.4 bar and an absolute pressure of 0.07 bar is maintained in the evaporator. Assuming that the properties of the solution are the same as those of water, and taking the overall heat transfer coefficient to be 2300 W m^{-2}K^{-1}, calculate the rate of steam consumption and the necessary heat transfer surface area.

The enthalpies of the vapour and liquor streams are a function of the pressure within the evaporator: h_V = 2572 kJ kg^{-1} (h_g at 0.07 bar) and h_L = 163kJ kg^{-1} (h_f at 0.07 bar). From steam tables: (if the steam and condensate remain saturated at 2.40 bar), h_S = 2715kJ kg^{-1} and h_C = 530kJ kg^{-1}. [**Ans:** S = 0.812 kg s^{-1} and A= 8.86m^2]

3. An aqueous solution at 15.5°C, and containing 4% solids, is concentrated to 20% solids. A single effect evaporator with a heat transfer surface area of 37.2 m^2 and an overall heat transfer coefficient of 2000 W m^{-2} K^{-1} is to be used. The calandria contains dry saturated steam at a pressure of 200 kPa and the evaporator operates under a vacuum of 81.3 kPa. If the evaporator temperature is 65.1 °C, calculate the evaporator capacity.

At 200 kPa the steam and condensate enthalpies are h_S = 2707 kJ kg^{-1}, h_C = 505 kJ kg^{-1}, T_s=120.2°C. Vapour enthalpy is 2618.6 kJ kg^{-1} and C_p is 1.91 kJ kg^{-1} K^{-1}. The enthalpy of the concentrated liquor stream at the evaporator temperature is h_L = 272 kJ kg^{-1} (h_f at 65.1°C).
[**Ans:** L = 0.393 kg s^{-1} and the evaporator capacity is F = 1.97 kg s^{-1}]

4. A single effect evaporator is used to evaporate orange juice at rate of 1000 Kg/h to concentrate from 10°Bx to 35°Bx. The temperature of feed is 20°C and boiling point is 82°C [Assume that boiling point remains constant]. The heat is supplied by steam where entrance and exit temperatures are 111°C and 82°C respectively. The specific heat capacity of steam and juice are 2.9 KJ/ Kg°C and 4.2 KJ/ Kg°C. The latent heat of vapour and steam are 2100 KJ/Kg and 2257 KJ/Kg respectively. Calculate the flow rate of steam, product, vapour and heat transfer area if overall heat transfer coefficient is 1700 W/m^2°C.

5. Grape juice at a rate of 3 kg/s is concentrated in a single effect evaporator from 18% to 23% solids content. Calculate the required heat transfer area of the evaporator if the juice enters the evaporator at 50°C, the juice boils in the evaporator at 50°C, saturated steam at 100°C is used as heating medium, the condensate exits at 100°C, the heat capacity of the juice is 3.7 kJ/kg°C and 3.6 kJ/kg°C at the inlet and the outlet of the evaporator respectively, and the overall heat transfer

coefficient is 1500 W/m²°C. The enthalpy of saturated steam, saturated vapour, and condensate is found from steam tables as: H_s at 100°C = 2676 kJ/kg; H_c at 100°C = 419 kJ/kg; π_s at 100°C = 2676 - 419 = 2257 kJ/kg; and, H_v at 50°C = 2592 kJ/kg. [**Ans: 20.8m²**]

6. Calculate the steam consumption in a single effect evaporator with 25 m² heat transfer area which is being used to concentrate a fruit juice. The juice enters the evaporator at 70°C, the saturation pressure in the evaporator is 31.19 kPa, the saturated steam at 100°C is used as the heating medium, the condensate exits at 95°C and the overall heat transfer coefficient is 1500 W/m² °C.

7. A single-effect evaporator is to produce a 35% solids tomato concentrate from a 6% solids raw juice entering at 18°C. The pressure in the evaporator is 20 kPa (absolute) and steam is available at 100 kPa gauge. The overall heat-transfer coefficient is 440 J m⁻² s⁻¹ °C⁻¹, the boiling temperature of the tomato juice under the conditions in the evaporator is 60°C, and the area of the heat-transfer surface of the evaporator is 12 m². Estimate the rate of raw juice feed that is required to supply the evaporator. [**Ans:** 536 kgh⁻¹]

8. Estimate (a) the evaporating temperature in each effect, (b) the requirements of steam, and (c) the area of heat transfer surface for a two effect evaporator. Steam is available at 100 kPa gauge pressure and the pressure in the second effect is 20 kPa absolute. Assume an overall heat-transfer coefficient of 600 and 450 J m⁻² s⁻¹ °C⁻¹ in the first and second effects respectively. The evaporator is to concentrate 15,000 kg h⁻¹ of raw milk from 9.5 % solids to 35% solids. Assume the sensible heat effects can be ignored, and that there is no boiling-point elevation. [**Ans:** (a) 1st. effect 94°C, 2nd. effect 60°C, (b) 5,746 kgh⁻¹, 0.53 kg steam/kg water (c) 450 m²]

9. A standard calandria type of evaporator with 100 tubes, each 1 m long, is used to evaporate fruit juice with approximately the same thermal properties as water. The pressure in the evaporator is 80 kPa absolute, and in the steam jacket 100 kPa absolute. Take the tube diameter as 5 cm. Estimate the rate of evaporation in the first evaporator. Assume the juice enters at 18°C and the overall heat transfer coefficient is 440 J m⁻² s⁻¹ °C⁻¹. [**Ans:** 72.5 kgh⁻¹]

Classification of evaporators on the basis of operation

Single effect evaporator

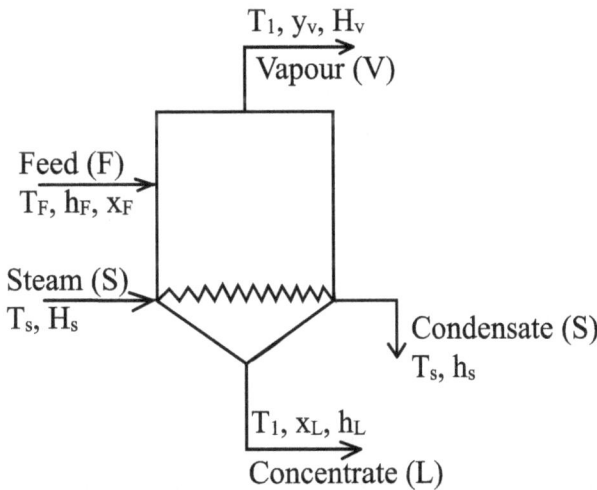

Fig. 4.8: Working of a single effect evaporator

The feed enters at T_F and saturated steam at T_s enters the heat exchange section. Condensed steam leaves as condensate or drips. Since the solution in the evaporator is assumed to be completely mixed, the concentrated product and the solution in the evaporator have same composition and temperature T_1 i.e. boiling point of solution. The temperature of vapor is also T_1. The pressure is P_1, which is vapor pressure of solution at T_1.

If feed solution is assumed to be dilute, then 1 kg of steam condensing will evaporate approximately 1 kg vapour. This will hold if the feed entering has temperature T_F near to boiling point T_1.

Rate of heat transfer, $q = UA(T_s - T_1) = UA \Delta T$

It is used for small capacity. The cost of steam is relatively cheap than evaporator cost.

Multiple effect evaporator

Multiple-effect evaporators are usually used. The theory is that if two evaporators are connected in series, the second effect can operate at a higher vacuum (and therefore at a lower temperature) than the first. The vapour evolved from the product in the first effect can be used as the heating medium for the next effect, which operates at a lower boiling temperature due to the higher vacuum. as shown in Fig. 4.9 and Fig. 4.10.

It is also possible to connect several evaporator effects in series to further improve steam economy. However, this makes the equipment more expensive and involves a higher temperature in the first effect. The total volume of product in the evaporator system increases with the number of effects connected in series. This is a drawback in the treatment of heat-sensitive products.

i. Forward feed

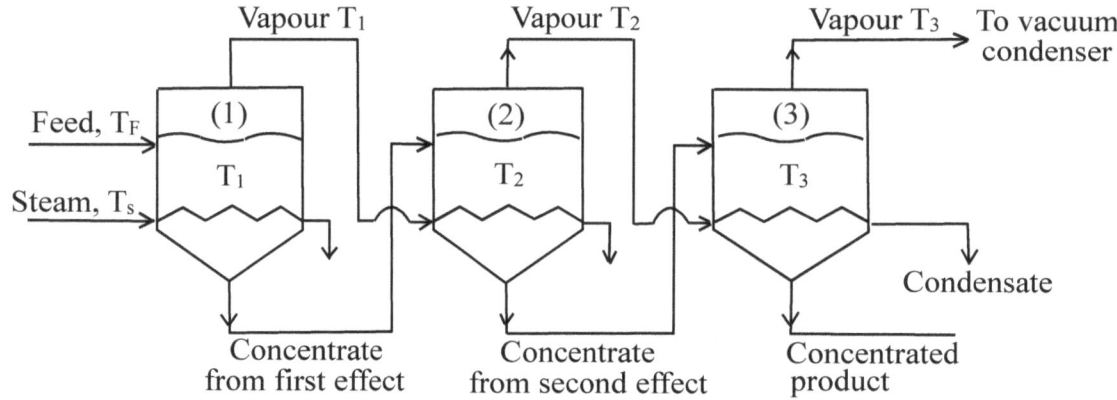

Fig. 4.9: Simplified diagram of forward-feed triple-effect evaporator

If the feed to the first effect is near the boiling point at pressure in the first effect, 1 kg of steam will evaporate almost 1 kg of water. The first operate at a temperature that is high enough that the evaporated water serves as a heating medium to the second and so on. Hence, the steam economy (kg vapour evaporated/ kg steam used) is increased.

However, the increased steam economy of a multiple effect evaporator is gained at a expense of original first cost of these evaporators.

This method of operation is used when the feed is hot or when the final concentrated product might be damaged at high temperature. The boiling point decreases from effect to effect. This means that if the 1st effect is at pressure (P_1= 1 atm), the last will be under vacuum pressure (P_3).

$P_1 > P_2 > P_3$ i.e. $T_1 > T_2 > T_3$.

Writing a heat balance for the first evaporator:

$$q_1 = U_1 A_1 (T_s - T_1) = U_1 A_1 \Delta T_1 \dots\dots\dots (4.8)$$

Similarly, in the second evaporator, remembering that the "steam" in the second is the vapour from the first evaporator and that this will condense at approximately the same temperature as it boiled, since pressure changes are small,

$$q_2 = U_2 A_2 (T_1 - T_2) = U_2 A_2 \Delta T_2 \dots\dots\dots (4.9)$$

If the evaporators are working in balance, then all of the vapours from the first effect are condensing and in their turn evaporating vapours in the second effect. Also assuming that heat losses can be neglected, there is no appreciable boiling-point elevation of the more concentrated solution, and the feed is supplied at its boiling point,

$$q_1 = q_2$$

Further, if the evaporators are so constructed that $A_1 = A_2$, the foregoing equations can be combined so that :

$$\frac{U_2}{U_1} = \frac{\Delta T_1}{\Delta T_2} \quad \ldots\ldots\ldots (4.10)$$

Equation 4.10 states that the temperature differences are inversely proportional to the overall heat transfer coefficients in the two effects. This analysis may be extended to any number of effects operated in series, in the same way.

ii. Backward feed

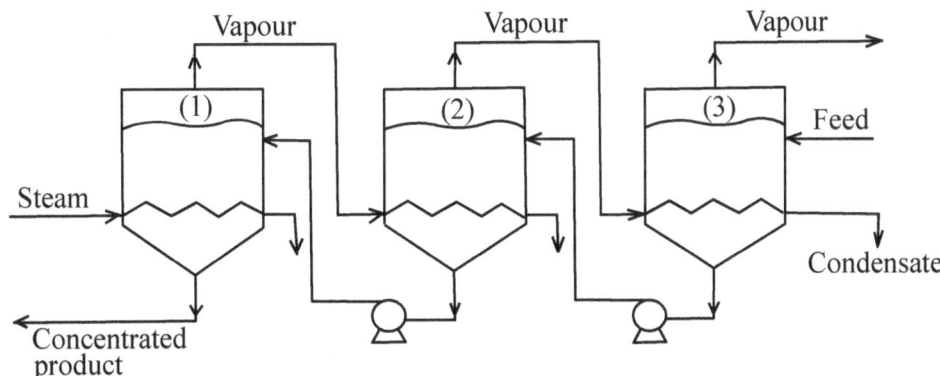

Fig. 4.10: Simplified diagram af backward-feed triple-effect evaporator

In the backward feed operation, the fresh feed enters the last and coldest effect and continues on until the concentrated product leaves the 1st effect. This method of reverse feed is advantageous when the fresh feed is cold, since a smaller amount of liquid must be heated to the higher temperature in 2nd and 1st effect. However, the liquid pump must be used in each effect, since the flow is from low to high pressure.

The reverse feed is used when the concentrated product is highly viscous. The high temperature in the early effects reduces the viscosity and gives reasonable heat transfer coefficients.

Advantages of multiple effect evaporators

At first sight, it may seem that the multiple effect evaporator has all the advantages, the heat is used over and over again and we appear to be getting the evaporation in the second and subsequent effects for nothing in terms of energy costs. Closer examination shows, however, that there is a price to be paid for the heat economy.

In the first effect, $q_1 = U_1 A_1 \Delta T_1$ and in the second effect, $q_2 = U_2 A_2 \Delta T_2$.

We shall now consider a single-effect evaporator, working under the same pressure as the first effect

$$q_s = U_s A_s \Delta T_s,$$ where subscript s indicates the single-effect evaporator.

Since the overall conditions are the same, $\Delta T_s = \Delta T_1 + \Delta T_2$, as the overall temperature drop is between the steam-condensing temperature in the first effect and the evaporating temperature in the second effect. Each successive steam chest in the multiple-effect evaporator condenses at the same temperature as that at which the previous effect is evaporating.

Now, consider the case in which $U_1 = U_2 = U_s$, and $A_1 = A_2$. The problem then becomes to find A_s for the single-effect evaporator that will evaporate the same quantity as the two effects. From the given conditions and from equation.,

$\Delta T_1 = \Delta T_2$

and, $\Delta T_s = \Delta T_1 + \Delta T_2 = 2 \Delta T_1$

Or, $\Delta T_1 = 0.5 \Delta T_s$

Now, $q_1 + q_2 = U_1 A_1 \Delta T_1 + U_2 A_2 \Delta T_2 = U_1(A_1 + A_2) \dfrac{\Delta T_s}{2}$

but $q_1 + q_2 = q_s$; and, $q_s = U A_s \Delta T_s$

so that, $\dfrac{(A_1 + A_2)}{2} = \dfrac{2A_1}{2} = A_s$

i.e. $A_1 = A_2 = A_s$

The analysis shows that if the same total quantity is to be evaporated, then the heat transfer surface of each of the two effects must be the same as that for a single effect evaporator working between the same overall conditions. The analysis can be extended to cover any number of effects and leads to the same conclusions. In multiple effect evaporators, steam economy has to be paid for by increased capital costs of the evaporators. Since the heat transfer areas are generally equal in the various effects and since in a sense what you are buying in an evaporator is suitable heat transfer surface, the n effects will cost approximately n times as much as a single effect.

Improving the economy of evaporators

The steam economy is the ratio of the amount of water evaporated to the amount of steam consumed. Poor evaporator economy may results from wasting the heat present in the vapours. Some of the techniques used to reclaim heat from the vapours include use of multiple effect evaporators, use of vapours to preheat the feed and vapour compression. The factors influencing the economy of evaporation are loss of concentrate and energy expenditure. Product losses are caused by:

a. Foaming, due to proteins and carbohydrates in the food, which reduce the rate of heat transfer and cause inefficient separation of vapour and concentrate, and

b. Entrainment, in which a fine mist of concentrate is produced during the violent boiling and is carried over the vapour.

Chapter 5: Distillation

Introduction

Distillation is a separation process, separating components in a mixture by making use of the fact that some components vaporize more readily than others. When vapours are produced from a mixture, they contain the components of the original mixture, but in proportions which are determined by the relative volatilities of these components. The vapour is richer in some components, those that are more volatile, and so a separation occurs.

In fractional distillation, the vapour is condensed and then re--evaporated when a further separation occurs. It is difficult and sometimes impossible to prepare pure components in this way, but a degree of separation can easily be attained if the volatilities are reasonably different. Where great purity is required, successive distillations may be used.

Major uses of distillation in the food industry are for concentrating essential oils, flavours and alcoholic beverages, and in the deodorization of fats and oils.

Distillation may be defined as the separation of the components of a liquid mixture by a process involving partial vapourization. The vapour evolved is usually recovered by condensation. This term is properly applied only to those operations where vapourization of a liquid mixture yields a vapour phase containing more than one constituent and it is desired to recover one or more of these constituents in the nearly pure state.

Ideal solution

An ideal solution is defined as the one in which there is complete uniformity in the cohesive forces. The solute- solute, solvent- solvent and solute- solvent interactions are identical. Thus, if there are two components A and B forming an ideal solution, then the intermolecular forces between A-A, A-B and B-B are essentially equal.

Raoult's, Dalton's and Henry's law

The partial pressure P_A of component A in a mixture of vapour is the pressure that would be exerted by component A at the same temperature, if present in the same volumetric concentration as in the mixture.

By Dalton's law of partial pressure, $P = \sum P_A$, i.e. the total pressure is equal to the summation of the partial pressures. Then, in an ideal mixtures, the partial pressure is proportional to the mole fraction of the constituent in the vapour phase, and:

$P_A = y_A P$ .... (5.1)

A simple relationship for the partial pressure developed by a liquid solute A in a solvent B is given by Henry's law in the form:

$P_A = H\, x_A$ .... (5.2)

Raoult's law states "The partial pressure of an ideal solution is dependent on the vapour pressure of each chemical component and the mole fraction of the component present in the solution".

$P_A = P_A^* \, x_A$ .... (5.3)

where,

P_A = partial pressure of component A in the solution.
P_A^* = vapour pressure of pure component A
x_A = mole fraction of component A in the solution.

Volatility and Relative volatility

Volatility

The volatility of any substance in a liquid solution may be defined as the equilibrium partial pressure of the substance in the vapour phase divided by the mole fraction of the substance in the liquid solution.

$$v_a = \text{volatility of component a in a liquid solution} = \frac{p_a}{x_a} \quad \text{.......... (5.4)}$$

The volatility of a material in the pure state is equal to the vapour pressure of the material in the pure state. Similarly, the volatility of a component in a liquid mixture which follows Raoult's law must be equal to the vapour pressure of that component in the pure state.

Relative volatility

Relative volatility (α), which is defined as the volatility of one component of a liquid mixture divided by the volatility of another component of the liquid mixture. Usually Relative volatilities are expressed with the higher of the two volatilities in the numerator.

$$\alpha_{ab} = \frac{v_a}{v_b} = \frac{p_a x_b}{x_a p_b}$$

If the vapours follow Dalton's law, $p_a = y_a P$ & $p_b = y_b P$: where P is the total pressure of the vapours,

Then,

$$\alpha_{ab} = \frac{y_a x_b}{y_b x_a} \quad \ldots\ldots\ldots (5.5)$$

This is often given as the definition of relative volatility, it can be calculated directly from vapour-liquid equilibrium data.

For binary mixtures:

$y_b = 1 - y_a$ & $x_b = 1 - x_a$, hence

$$y_a = \frac{\alpha x_a}{1 + (\alpha - 1)x_a} \quad \ldots\ldots\ldots (5.6)$$

$$x_a = \frac{y_a}{\alpha + (1 - \alpha)y_a} \quad \ldots\ldots\ldots (5.7)$$

Boiling point diagram

The composition of the vapour in equilibrium with a liquid of given composition is determined experimentally using an equilibrium still. The results are conveniently shown on temperature-composition diagram as shown in Fig. 5.1.

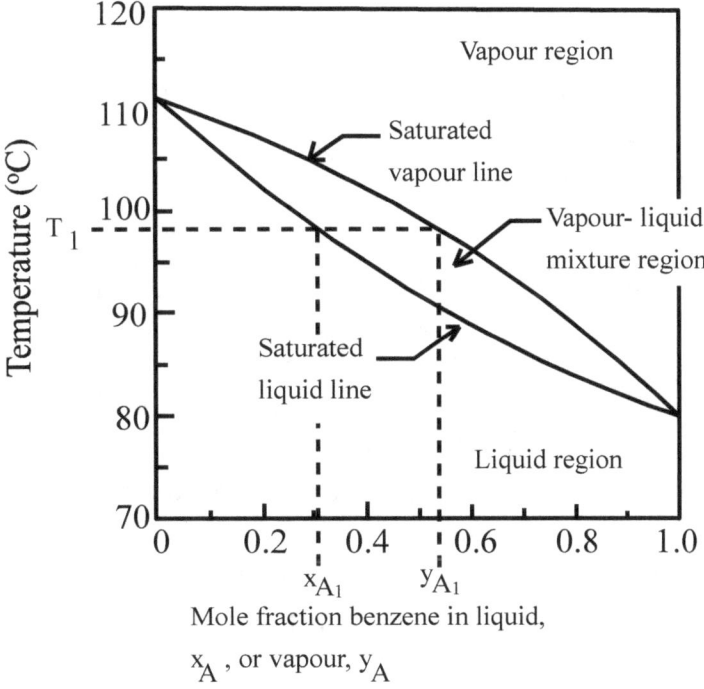

Fig. 5.1: Boiling point diagram of benzene- toluene mixture at 1 atm

The vapour- liquid equilibrium relations of a binary mixture are given as the boiling point diagram as shown in the Fig. 5.1. In the Fig. 5.1, the vapour liquid equilibrium relations of benzene (A) and toluene (B) at 1 atm pressure is shown.

The upper line is the saturated vapour line or the dew point line and the lower line is the saturated liquid line or the bubble point line. The two phase region is the region between these two lines.

For example, if we heat a cold liquid mixture of benzene and toluene ($x_{A1} = 0.318$ benzene) and heat the mixture it will start to boil at 98°C and the composition of the first vapour in equilibrium is $y_{A1} = 0.532$. As we continue boiling, the composition x_A will move to the left since y_A is richer in A.

The system benzene- toluene follows Raoult's law so the boiling point diagram can be calculated from the following equation:

$$P_A + P_B = P$$
$$P_A*(x_A) + P_B*(1-x_A) = P$$
$$y_A = \frac{P_A}{P} = \frac{P_A * x_A}{P} \ldots\ldots\ldots(5.8)$$

Generally, the equilibrium data are plotted as y_A Vs x_A and a 45° line is given to show that y_A is richer in component A than x_A.

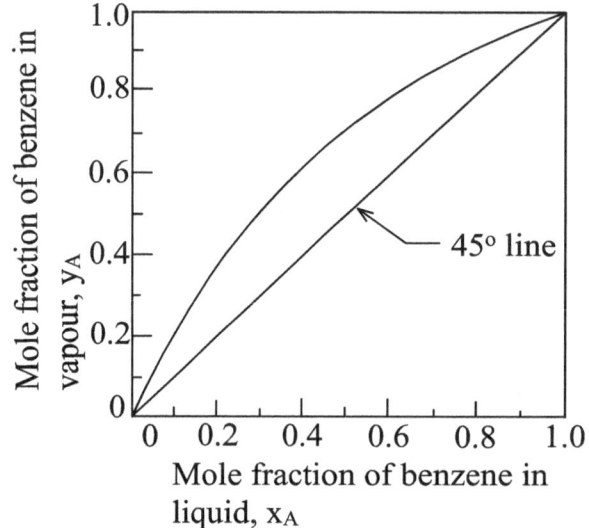

Fig. 5.2: **Equilibrium diagram for ideal solution at 1 atm (zeotropic)**

Eg: benzene- toluene, n- heptanes- toluene, carbon disulphide- carbon tetra chloride.

Boiling point diagram for non-ideal solutions

The boiling point diagram differs considerably.

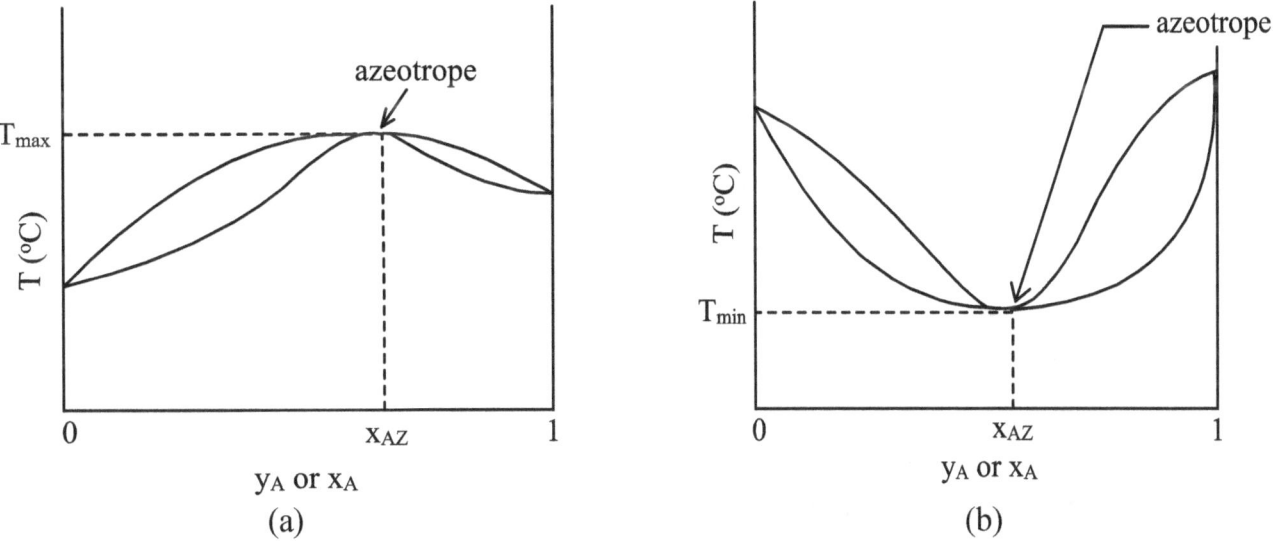

**Fig. 5.3: (a) Maximum boiling azeotrope (eg: Acetone- Chloroform) and
(b) Minimum boiling azeotrope (eg: Ethanol- water)**

In both Fig. 5.3 (a) and (b), there is a critical composition xg where the vapour has the same composition as the liquid, so that no change occurs on boiling. Such critical mixtures are called azeotropes, and special methods are necessary to effect separation. For compositions other than x_{AZ}, the vapour formed will have a different composition from that of the liquid. It is important to note that these diagrams are for constant pressure conditions, and that the composition of the vapour in equilibrium with a given liquid will change with pressure.

Methods of Distillation

There are 3 main methods used in distillation practice which all rely on its basic fact:
i. Differential distillation
ii. Flash or Equilibrium distillation, and
iii. Rectification

Differential (Batch) distillation

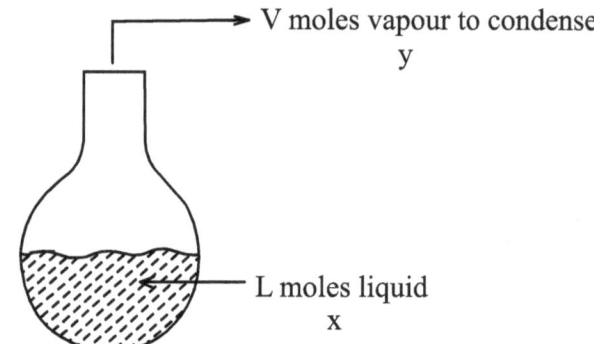

Fig. 5.4: Differential distillation

1. Liquid is charged to a heated kettle
2. The liquid charge is boiled slowly
3. The vapours are withdrawn as quickly as they form to a condenser
4. The condensed vapour (distillate) is collected.

    The first portion of vapour condensed will be richest in the more volatile component A. As vapourization proceeds, the vapourized product becomes leaner in A. i.e., the composition changes with time.

    The total mole amount in the liquid is L with the mol fraction of A being x. Assume a small amount of dL is vapourized so that the composition of the liquid changes from x to (x-dx) and the amount of liquid from L to (LdL). Let y be the composition of A in the vapour.

The material balance of A gives

$xL = (x-dx)(L-dL) + ydL$ .... **(5.9)**

Expanding the right side, we have

$xL = xL - xdL - Ldx + dxdL + ydL$ .... **(5.10)**

Ignoring the double derivative term dxdL and rearranging,

$$\frac{dL}{L} = \frac{dx}{(y-x)} \quad \text{.......... (5.11)}$$

Integrating,

$$\int_{L_1}^{L_2} \frac{dL}{L} = \ln \frac{L_1}{L_2} = \int_{x_2}^{x_1} \frac{dx}{y-x} \quad \text{.......... (5.12)}$$

where L_1 is the original moles charged, L_2 the moles left in the still, x_1 the original composition, and x_2 the final composition of liquid. Equation 5.12 is known as the Rayleigh equation.

The equilibrium curve gives the relationship between y and x. Then the integration of Rayleigh equation can be done numerically or graphically between x_1 and x_2.

The average composition of total material distilled, y_{av}, can be obtained using the material balance:

$$L_1 x_1 = L_2 x_2 + V\, y_{av} \quad \ldots\ldots\ldots\ (5.13)$$

$$V = L_1 - L_2 = \text{moles distilled} \quad \ldots\ldots\ldots\ (5.14)$$

Flash or equilibrium distillation

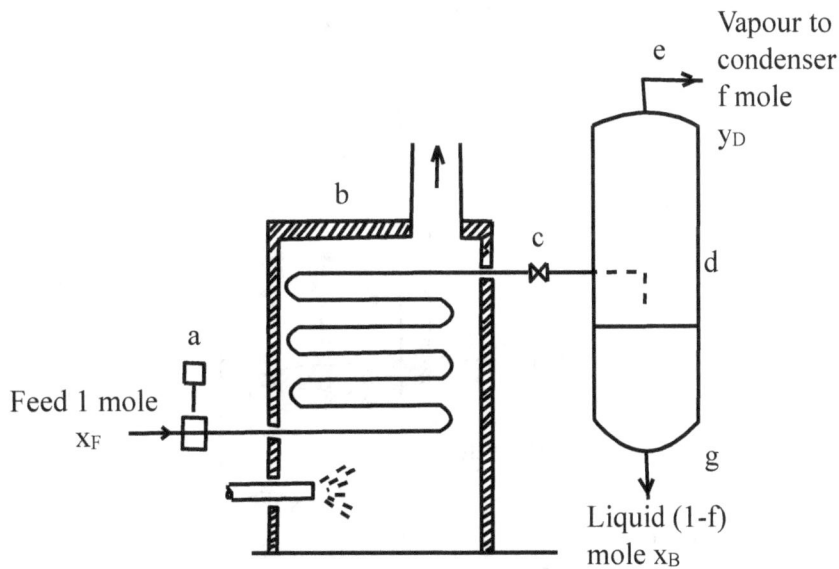

Fig. 5.5: Flash distillation

Flash distillation vapourizes a definite fraction of the liquid, the evolved vapour is in equilibrium with the residual liquid, the vapour is separated from the liquid and condensed.

Consider 1 mole of a binary mixture fed to the above equipment. By a material balance for the more volatile component, we have;

$$x_F = f y_D + (1-f) x_B \quad \ldots\ldots\ldots\ (5.15)$$

where

x_F = concentration (mole fraction) of A in the feed

y_D and x_B = concentrations of A in the vapour and liquid

$f = \dfrac{V}{F}$ = the molal fraction of the feed to be vaporized

V = moles per hour of vapour

F = moles per hour of feed

$L = F - V$ = moles per hour of liquid

Both y_D and x_B are unknown, but they are on the equilibrium curve.

In general we have the following operating equation for flash distillation by rearranging equation 5.15;

$$y = \frac{f-1}{f}x + \frac{x_F}{f} \quad \ldots\ldots\ldots (5.16)$$, which passes the point (x, x_F).

Rectification

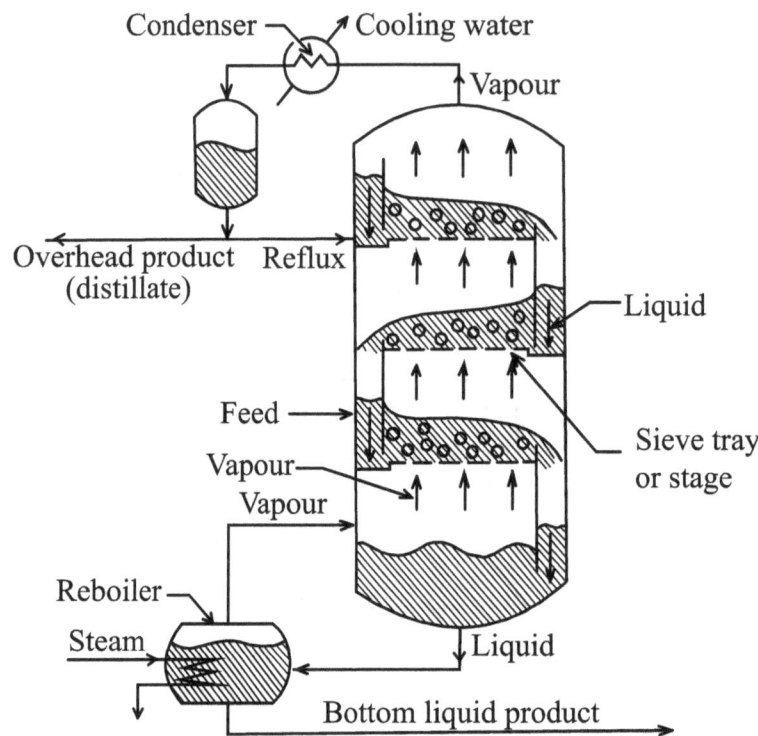

Fig. 5.6: Working of rectification column

In the two processes outlined previously, the vapour leaving the still at any time is in equilibrium with the liquid remaining, and there will normally be only a small increase in concentration of the more volatile component. The essential merit of rectification is that it enables a vapour to be obtained that is substantially richer than the liquid left in the still.

This is achieved by an arrangement known as rectification column, which enables successive vaporization and condensation to be accomplished in one unit as shown in Fig. 5.6. The feed enters the column somewhere in the middle of the column. If the feed is liquid, it flows down to sieve tray or stage. Vapour enters the tray and bubbles through the liquid on this tray as the entering liquid flows across. The vapour continues up to the next tray or stage, where it is again contacted with the down flowing liquid.

In this case, the concentration of the more volatile component (the lower boiling component A) is being increased in the vapour from each stage going upward and decreased in the liquid from each stage going downward. The final vapour product coming overhead is condensed in a condenser.

The portion of the liquid product (distillate) is removed, which contains a high concentration of A. The remaining liquid from the condenser is returned (refluxed) as a liquid to the top tray. The liquid leaving the bottom tray enters a reboiler, where it is partially vapourized and the remaining liquid, which is lean in A or rich in B, is withdrawn as liquid product.

Equations of operating lines

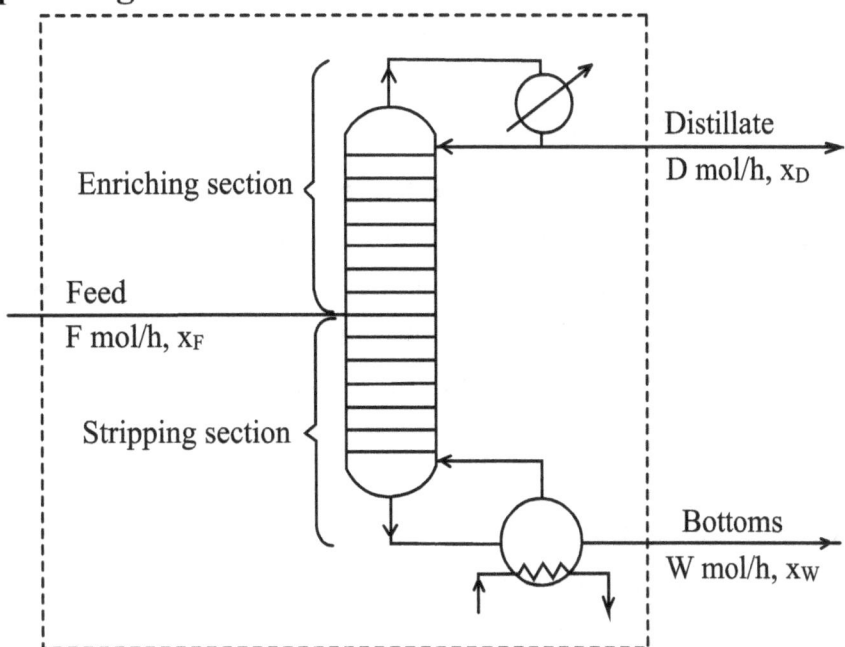

Fig. 5.7: Continuous rectification column

In Fig. 5.7, a continuous column is shown with feed being introduced to the column at an intermediate point and an overhead distillate product and the bottom product being withdrawn. The upper part of the tower above the feed entrance is called enriching section, since the entering feed of binary components A & B is enriched in this section, so that the distillate is richer in A than the feed. The tower is at steady state.

An overall material balance around the entire column is:

$F = D + W$ .... **(5.17)**

The total material balance on component A gives:

$F \cdot x_F = D \cdot x_D + W \cdot x_W$ .... **(5.18)**

Equation for enriching section

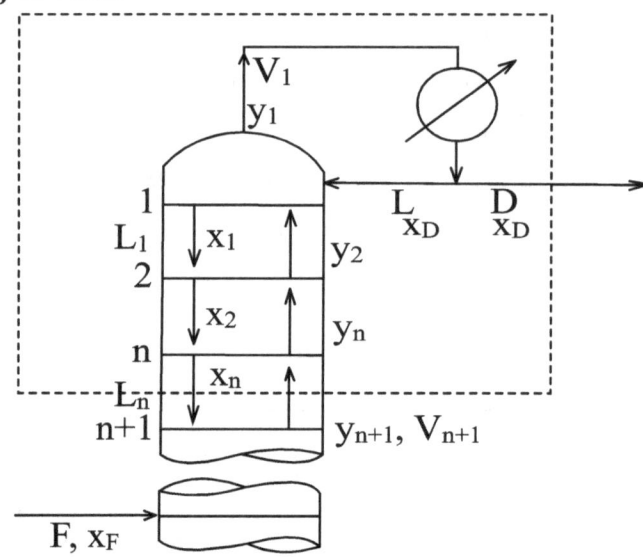

Fig. 5.8: Enriching section of a rectification column

The above Fig. 5.8 shows the enriching section above the feed. The vapour from top tray having a composition y_1 passes to the condenser where it is condensed so that the resulting liquid is at the boiling point. The reflux stream L mole/hr and distillate D mole/hr have same composition so that $y_1 = x_D$. Since equimolar overflow is assumed., $L_1 = L_2 = L_n$ & $V_1 = V_2 = V_{n+1}$.

Making a total material balance over the dashed line section in the above Fig. 5.8:

$$V_{n+1} = L_n + D$$

Making a balance on Component A:

$$V_{n+1} \cdot y_{n+1} = L_n \cdot x_n + D \cdot x_D$$

$$\text{Or, } y_{n+1} = \frac{L_n \cdot x_n}{V_{n+1}} + \frac{D \cdot x_D}{V_{n+1}}$$

Since, $V_{n+1} = L_n + D$, $\dfrac{L_n}{V_{n+1}} = \dfrac{R}{R+1}$ and hence above equation becomes:

$$y_{n+1} = \frac{R \cdot x_n}{(R+1)} + \frac{x_D}{(R+1)} \quad \ldots\ldots\ldots (5.19)$$

where $R = \dfrac{L_n}{D}$ = Reflux ratio = Constant

The above equation is a straight line on a plot of vapour composition Vs liquid composition. The slope of line is L_n/V_{n+1} or $R/(R+1)$. It intersects the y=x line at x= x_D. The intercept of the operating line at x =0 is y = x_D /(R+1).

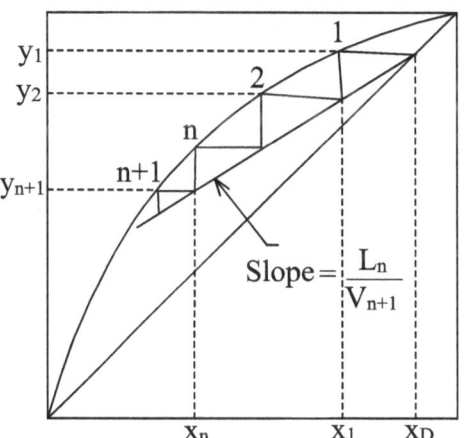

The theoretical stages are determined by starting at the operating line at x_D and moving horizontally to intersect the equilibrium line at x_1. Then y_2 is the composition of the vapour passing the liquid x_1. Similarly, other theoretical trays are stepped off down the tower in the enriching section to the feed tray.

Equation for stripping section

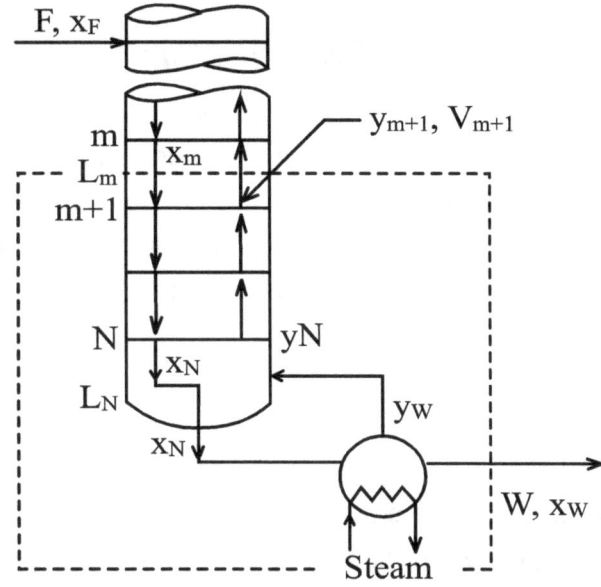

Fig. 5.9: Stripping section of a rectification column

Making a total material balance over a dashed line in above Fig. 5.9 for stripping section below feed entrance:

$L_m = V_{m+1} + W$

Or, $V_{m+1} = L_m - W$

103

Making a balance on component A:

$V_{m+1} \cdot y_{m+1} = L_m \cdot x_m - W \cdot x_w$

Or, $y_{m+1} = \dfrac{L_m \cdot x_m}{V_{m+1}} - \dfrac{W \cdot x_w}{V_{m+1}}$ .... **(5.20)**

Since, equimolar flow is assumed, $L_m = L_N$ = Constant and $V_{m+1} = V_{m+2} = V_N$ = constant.

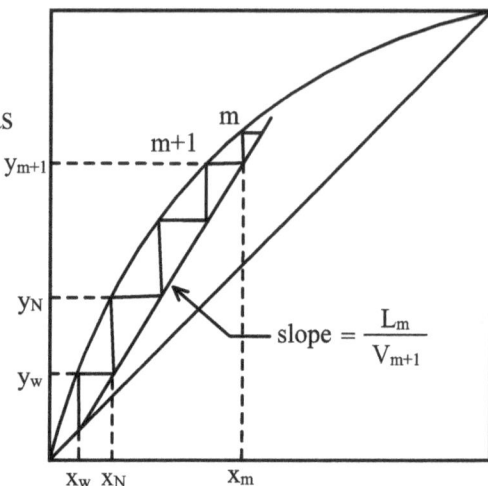

The above equation is a straight line when plotted as y Vs x in boiling point diagram with a slope of L_m / V_{m+1}. It intersects the y = x line at x = x_w. The intercept at x=0 is $y = -W \cdot x_w / V_{m+1}$.

Again the theoretical stages for the stripping section are determined by starting at x_w, going up to intersect the equilibrium line at at y_N, and so on.

Exercises on operating sections of rectification

1. The feed of 100 kg mol/hr is liquid containing 45% benzene and 55 mol% toluene and enters at the boiling point. A distillate containing 95 mol% benzene and 5 mol% toluene and bottom containing 10 mol% benzene and 90 mol% toluene are to be obtained. The reflux ration is 4:1. The average heat capacity of the feed is 159 KJ/ Kg moleK and the average latent heat is 32099 KJ/kgmol. Calculate the operating line for enriching and stripping section. [**Ans:** $y_{n+1} = 0.8\, x_n + 0.190$ & $y_{m+1} = 1.285\, x_m - 0.02857$]

Effect of feed conditions

The condition of the feed stream F entering the tower determines the relation between the vapour V_m in the stripping section and V_n in the enriching section as well as between L_m and L_n. If the feed is part liquid and part vapour, the vapour will add to V_m to give V_n, and the liquid will add to L_n to give L_m. We represent the condition of the feed by quantity 'q' which defines as:

$$q = \dfrac{\text{Heat needed to vaporize 1 mol of feed at entering conditions}}{\text{Molar latent heat of vaporization of feed}}$$

$$= \dfrac{H_v - H_F}{H_v - H_L} = \dfrac{(H_v - H_L) + (H_L - H_F)}{H_v - H_L} = 1 + \left(\dfrac{H_L - H_F}{H_v - H_L}\right) \quad \ldots \ldots \ldots (5.21)$$

Where H_V is the enthalpy of the feed at the dew point, H_L the enthalpy of the feed at the boiling point (bubble point), and H_F the enthalpy of the feed at its entrance conditions.

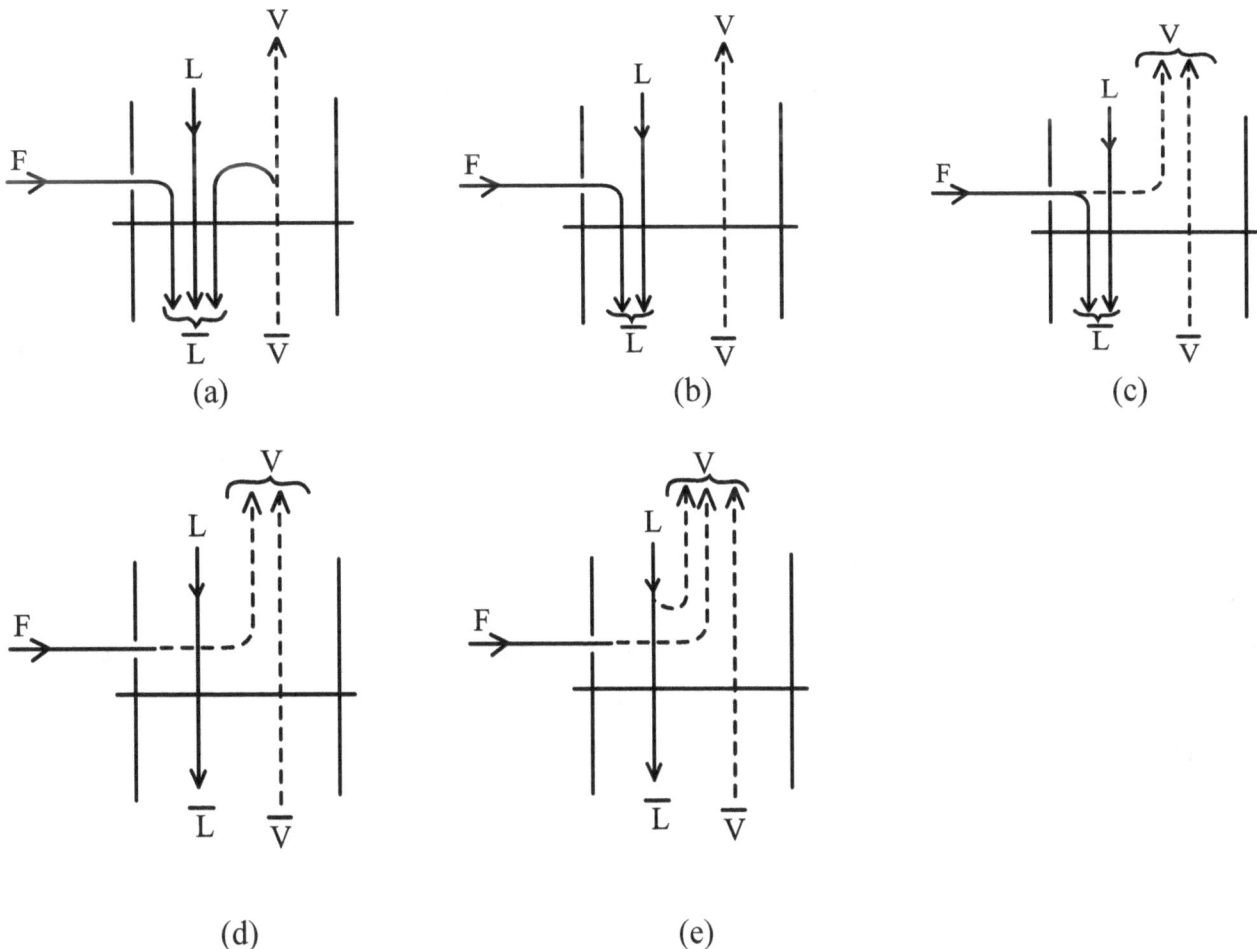

Fig. 5. 10: Effect of feed conditions (a) cold liquid [q>1] (b) at bubble point or saturated liquid [q=1] (c) partial vapour [0<q<1] (d) at dew point or saturated vapour [q=0] (e) superheated vapour [q<0]

Equation of q- line

'q' can also be considered as the number of moles of saturated liquid produced on the feed plate by each mole of feed added to the tower.

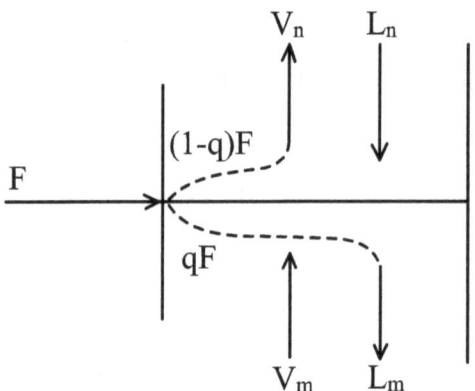

Fig. 5.11: Entrance of feed to the rectification column

$L_m = L_n + qF$ .... (5.22)

$V_n = V_m + (1 - q) F$ .... (5.23)

The point of intersection of the enriching and stripping operating line equations on the y-x plot can be derived as follows:

$V_n.y = L_n.x + D.x_D$ .... (5.24)

$V_m.y = L_m.x - W.x_w$ .... (5.25)

Where the y and x values are the point of intersection of the two operating lines. (5.24) & (5.25) gives:

$(V_m - V_n) y = (L_m - L_n) x - (Dx_D + Wx_w)$ .... (5.26)

We know that:

$F.x_F = D.x_D + W.x_W$ .... (5.27)

Substituting equation (5.22), (5.23) & (5.27) in equation (5.26) and rearranging:

$$\boxed{y = \frac{q}{q-1}.x - \frac{x_F}{q-1}}$$ .... (5.28)

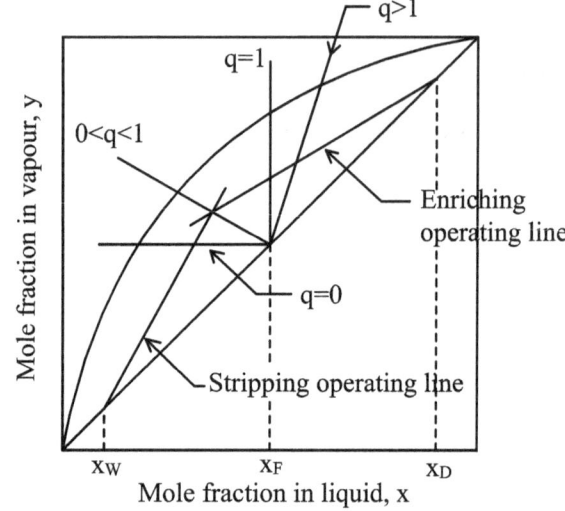

The q-line equation passes the intersection of the two operating lines. The q-line has a slope of q/(q-1) and passes through the 45° line at $y = x = x_F$. The above Fig. is an example of a feed of part liquid and part vapour (0<q<1). First draw the q-line, then the enriching line, the intercept of these two lines can be used to draw the stripping line.

Exercises on q-line equation

1. 100 Kmol/hr of a feed at 60% benzene and 40% heptane is to be separated by distillation. The distillate is to be 90% benzene and 10% bottom of benzene. The feed enters the column as 30 mol% vapour. Use R 3 times the minimum. Assume a constant relative volatility of α of and the pressure is constant through the column at 1 atm. Obtain the equations of the q- line. [**Ans:** $y= -2.333x +2$]

Location of the feed tray in a tower and number of ideal plates (McCabe-Thiele method)

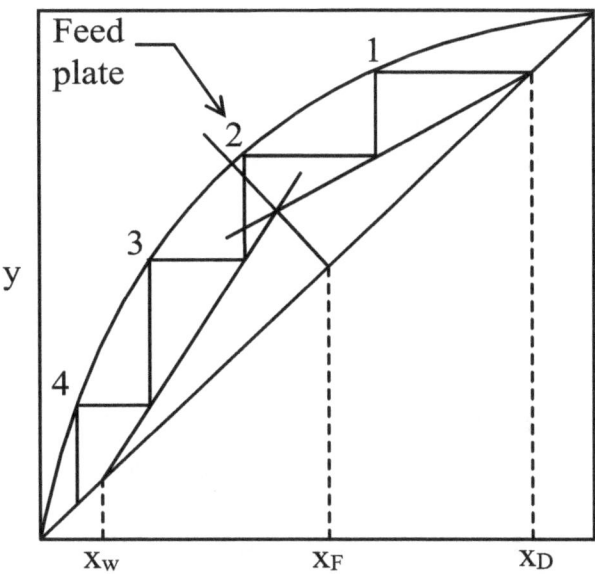

Fig. 5.12: McCabe thiele method for counting number of plates in a rectification column

To determine the number of theoretical trays needed in a tower, the stripping and enriching lines are drawn to intersect on the q line as follows:

i. Starting from the top at x_D, the trays are stepped off along the enriching line.

ii. It is important to switch to the stripping line when the triangle first passes the q line (intersection of the 2 operating lines), which is tray 2 in this case.

iii. The trays are continued to step off along the stripping line.

iv. The number of theoretical trays required is 3.7 with the feed on tray 2.

v. In the above Fig. 5.12, the feed is part liquid and part vapour, since $0<q<1$. Hence, in adding the feed to tray 2, the vapour portion of the feed is separated and added to below plate 2 and the liquid added to the liquid from above entering tray 2.

vi. If the feed is all liquid, it should be added to the liquid flowing to tray 2 from the tray above. If the feed is all vapour, it should be added below tray 2 and join in the vapour rising from the plate below.

vii. Since a reboiler is considered as a theoretical step, when the vapour y_B is in equilibrium with x_B, the number of theoretical trays in a tower is equal to the number of theoretical steps minus one. The value of q for cold liquid feed is:

$$q = 1 + \frac{c_{pL}(T_B - T_F)}{H_v - H_L}$$

The value of q for superheated vapour feed is:

$$q = 0 + \frac{c_{pV}(T_D - T_F)}{H_v - H_L}$$

where

c_{pL} = specific heat capacity of feed liquid
c_{pV} = specific heat capacity of feed vapour
T_F = temperature of feed
T_B = bubble point of feed
T_D = dew point of feed

Example 5.1: Number of plates in rectification column

1. From the previous numerical data (Page 104 exercise), determine the number of theoretical plates needed in the rectification column.

T (K)	353.3	358.2	363.2	368.2	373.2	378.2	383.8
x_A	1.0	0.78	0.581	0.411	0.258	0.130	0
y_A	1.0	0.90	0.777	0.632	0.456	0.261	0

Sol^n:
Given,
F = 100 kmol/h, x_F = 0.45, x_D = 0.95, x_B = 0.1,
R = L_n /D = 4
Overall material balance from equation 5.17;
F = D + B
100 = D + B
Benzene balance using equation 5.18;

$F \cdot x_F = D \cdot x_D + B \cdot x_B$

Or, $100 (0.45) = D (0.95) + (100-D)(0.10)$

Hence, D = 41.2 kmol/h and B = 58.8 Kmol/h

Using equation 5.19, the enriching operating line is

$$y_{n+1} = \frac{R}{R+1} x_n + \frac{x_D}{R+1} = \frac{4}{4+1} x_n + \frac{0.95}{4+1} = 0.80 x_n + 0.19$$

Using equation 5.28, the q- line is

$$y = \frac{q}{q-1} x - \frac{x_F}{q-1}$$

$$q = 1 + \frac{H_L - H_F}{H_v - H_L}$$

The value of $H_v - H_L$ = Latent heat = 32099 kJ/ kmol

$H_L - H_F = C_{pL}(T_B - T_F)$

Where the heat capacity of the liquid feed C_{pL} = 159 kJ/ kmol K, T_B = 366.7 K (Boiling point of feed), and T_F = 327.6K (Inlet feed temperature).

$$q = 1 + \frac{C_{pL}(T_B - T_F)}{H_v - H_L} = 1 + \frac{159(366.7 - 327.6)}{32099} = 1.195$$

So the q - line is:

$$y = \frac{1.195}{0.195} x - \frac{x_F}{0.195} = 6.128 x - 5.128 x_F$$

The enriching and q lines are plotted in the figure. Their intersection identifies one point in the stripping line. Linking this point to the bottom point $y = x = x_B = 0.1$, we obtain the stripping line. The number of theoretical steps is 7.6, or 7.6 steps minus a reboiler, which gives 6.6 theoretical trays. The feed is introduced on tray 5 from the top.

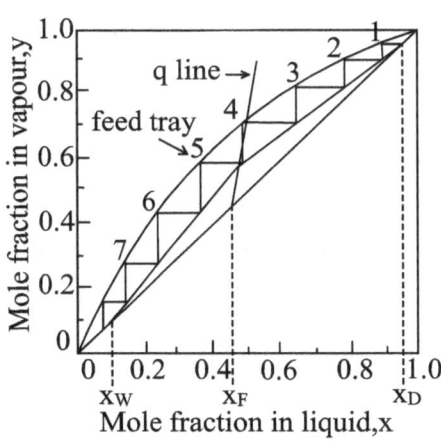

Exercises on calculating number of plates in rectification

1. A distillation column is fed with a mixture of benzene and toluene, in which the mole fraction of benzene is 0.35. The column is to yield a product in which the mole fraction of benzene is 0.95, when working with a reflux ratio of 3.2, and the waste from the column is not to exceed 0.05 mole fraction of benzene. If the plate efficiency is 60 per cent, estimate the number of plates required and the position of the feed point. The relation between the mole fraction of benzene in liquid and in vapour is given by:

x_A	0.1	0.2	0.3	0.4	0.5	0.6	0.7	0.8	0.9
y_A	0.20	0.38	0.51	0.63	0.71	0.78	0.85	0.91	0.96

[**Ans:** Theoretical plate = 10, The feed tray lies between ideal trays 5 and 6]

2. The relationship between the mole fraction of carbon disulphide in the liquid and in the vapour during the distillation of a carbon disulphide–carbon tetrachloride mixture is:

x_A	0	0.20	0.40	0.60	0.80	1.00
y_A	0	0.445	0.65	0.795	0.91	1.00

Determine graphically the theoretical number of plates required for the rectifying and stripping portions of the column. The reflux ratio = 3, the slope of the fractionating line = 1.4, the purity of product = 99 per cent, and the concentration of carbon disulphide in the waste liquors = 1 per cent. What is the minimum slope of the rectifying line in this case? [**Ans :** Plate = 12]

3. A mixture of benzene and toluene containing 40 mole per cent benzene is to be separated to give a product containing 90 mole per cent benzene at the top, and a bottom product containing not more than 10 mole per cent benzene. The feed enters the column at its boiling point, and the vapour leaving the column which is condensed but not cooled, provides reflux and product. It is proposed to operate the unit with a reflux ratio of 3 kmol/kmol product. It is required to find the equations of operating lines, the number of theoretical plates needed and the position of entry for the feed. [**Ans:** $y_n = 0.75 x_{n+1} + 0.225$, $y_m = 1.415 x_{m+1} - 0.042$, plates =7]

4. A continuous rectifying column handles a mixture consisting of 40 per cent of benzene by mass and 60 per cent of toluene at the rate of 4 kg/s, and separates it into a product containing 97 per cent of benzene and a liquid containing 98 per cent toluene. The feed is liquid at its boiling-point.

x_A	0.1	0.2	0.3	0.4	0.5	0.6	0.7	0.8	0.9
y_A	0.22	0.38	0.51	0.63	0.70	0.78	0.85	0.91	0.96

(a) Calculate the mass flows of distillate and waste liquor.

(b) If a reflux ratio of 3.5 is employed, how many plates are required in the rectifying part of the column?

(c) What is the actual number of plates if the plate-efficiency is 60 per cent? [**Ans:** 1.6 &2.4 kg/s; 10 &17]

Total reflux ratio

In distillation of a binary mixture A and B, the feed conditions, distillate and bottoms compositions are usually specified and the number of theoretical trays are to be calculated. The number of theoretical trays depends on the operating lines. To fix the operating lines, the reflux ratio $R = L_n/D$ at the top must be set.

One limiting case is total reflux, $R = \infty$, or $D = 0$. The material balance becomes

$$V_{n+1} = L_n \quad \text{......... (5.29)}$$

Or, $V_{n+1}y_{n+1} = L_n x_n$ .... (5.30)

Hence, the operating lines of both sections are on the 45° line,

$$y_{n+1} = x_n \quad \text{......... (5.31)}$$

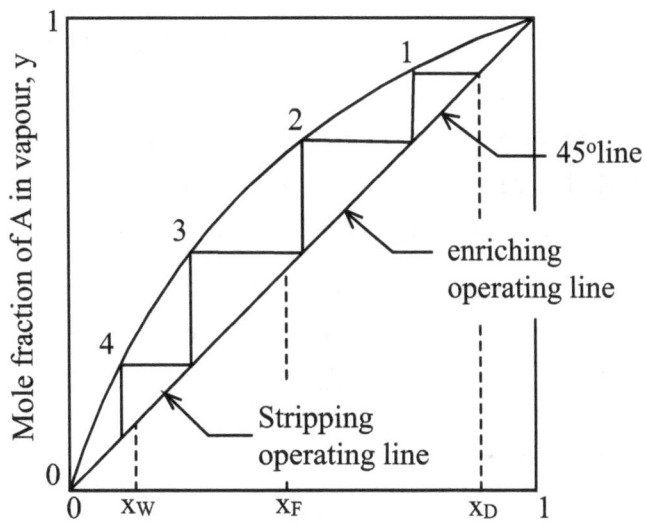

Fig. 5.13: Case of total reflux

Total reflux is an extreme case, the number of theoretical trays required is at its minimum to obtain the given separation of x_D and x_B as given in Fig. 5.13. However, in reality we have no product at all, and the tower diameter is infinite.

Minimum reflux ratio

Another limiting case is the minimum reflux ratio, R_m, that will require infinite number of trays for the given separation of x_D and x_B. This corresponds to the minimum vapour flow in the tower, and hence the minimum reboiler and condenser sizes as of Fig. 5.14.

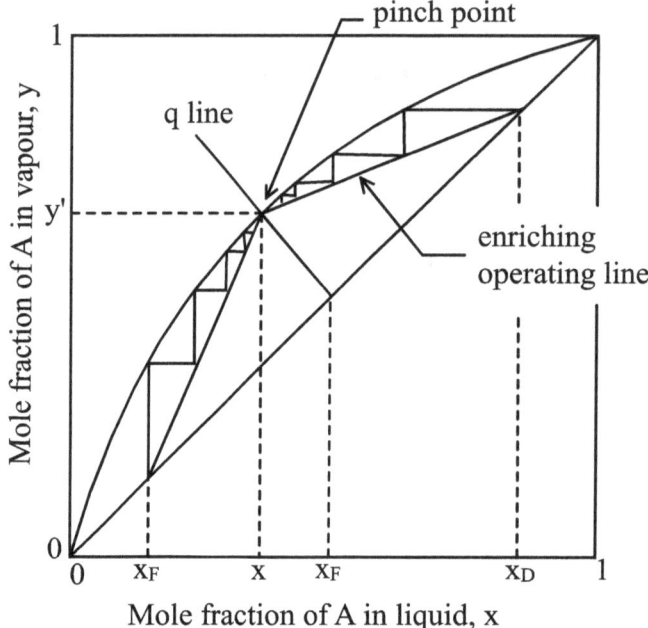

Fig. 5.14: Case of minimum reflux ratio

If the reflux ratio R decreases, the slope of the enriching line R/(R+1) decreases, the intersection of this line and the stripping line with the q line moves farther from the 45° line and closer to the equilibrium line. The number of steps required to give a fixed x_D and x_w increases. At the extreme case, the two operating lines touch the equilibrium line, a "pinch point" at y' and x' occurs, where the number of steps required is infinite. The slope of enriching line in this case is:

$$\frac{R_m}{R_m+1} = \frac{x_D - y'}{x_D - x'} \quad \ldots\ldots\ldots (5.32)$$

Chapter 6: Drying

Introduction

Drying is one of the oldest methods of preserving food. Primitive societies practiced the drying of meat and fish in the sun long before recorded history. Today the drying of foods is still important as a method of preservation. Dried foods can be stored for long periods without deterioration occurring. The principal reasons for this are that the microorganisms which cause food spoilage and decay are unable to grow and multiply in the absence of sufficient water and many of the enzymes which promote undesired changes in the chemical composition of the food cannot function without water.

Preservation is the principal reason for drying, but drying can also occur in conjunction with other processing. For example in the baking of bread, application of heat expands gases, changes the structure of the protein and starch and dries the loaf.

Losses of moisture may also occur when they are not desired, for example during curing of cheese and in the fresh or frozen storage of meat, and in innumerable other moist food products during holding in air.

Drying of foods implies the removal of water from the foodstuff. In most cases, drying is accomplished by vapourizing the water that is contained in the food, and to do this the latent heat of vapourization must be supplied. There are, thus, two important process-controlling factors that enter into the unit operation of drying:

(a) Transfer of heat to provide the necessary latent heat of vapourization,
(b) Movement of water or water vapour through the food material and then away from it to effect separation of water from foodstuff.

Drying processes fall into three categories:

a. *Air and contact drying under atmospheric pressure*: In air and contact drying, heat is transferred through the foodstuff either from heated air or from heated surfaces. The water vapour is removed with the air.

b. *Vacuum drying*: In vacuum drying, advantage is taken of the fact that evaporation of water occurs more readily at lower pressures than at higher ones. Heat transfer in vacuum drying is generally by conduction, sometimes by radiation.

c. *Freeze drying*: In freeze drying, the water vapour is sublimed off frozen food. The food structure is better maintained under these conditions. Suitable temperatures and pressures must be established in the dryer to ensure that sublimation occurs.

Equilibrium moisture content (EMC)

The equilibrium vapour pressure above a food is determined not only by the temperature but also by the water content of the food, by the way in which the water is bound in the food, and by the presence of any constituents soluble in water. Under a given vapour pressure of water in the surrounding air, a food attains a moisture content in equilibrium with its surroundings when there is no exchange of water between the food and its surroundings. This is called its equilibrium moisture content.

It is possible, therefore, to plot the equilibrium vapour pressure against moisture content or to plot the relative humidity of the air in equilibrium with the food against moisture content of the food. Often, instead of the relative humidity, the water activity of the food surface is used. This is the ratio of the partial pressure of water in the food to the vapour pressure of water at the same temperature.

The equilibrium curves obtained vary with different types of foodstuffs and examples are shown in Fig. 6.1 below.

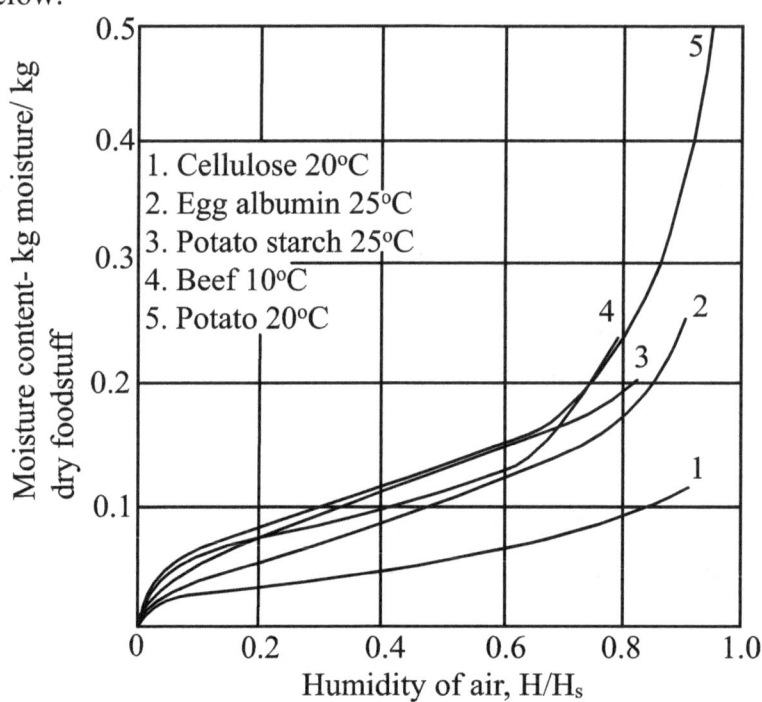

Fig. 6.1: Equilibrium moisture contents

Thus, for the potato starch as shown in Fig. 6.1, at a temperature of 25°C in an atmosphere of relative humidity 30% (giving a water activity of 0.3), the equilibrium moisture content is seen to be 0.1 kg water/kg dry potato. It would not be possible to dry potato starch below 10% using an air dryer with air at 25°C and relative humidity 30 %. It will be noted from the shape of the curve that above a certain relative humidity, about 80% in the case of potato starch, the equilibrium content increases very rapidly with increase in relative humidity. There are marked differences between foods, both in shape of the curves and in the amount of water present at any relative humidity and temperature, in the range of relative humidity between 0 and 65 %. The sigmoid (S-shaped) character of the curve is most pronounced, and the moisture content at low humidities is greatest, for food whose dry solids are high in protein, starch, or other high molecular weight polymers. They are low for foods high in soluble solids. Fats and crystalline salts and sugars, in general absorb negligible amounts of water when the RH is low or moderate. Sugars in the amorphous form absorb more than in the crystalline form.

Moisture content representations

The moisture content of a material may be expressed in a wet- weight basis (wb), i.e. Kg of water per Kg of wet material, or in a dry- weight basis (db), i.e. Kg of water per Kg of dry solids. The latter method is more commonly used in drying calculations.

Wet basis (wb):

The moisture content in this method is represented by the following expression:

$$m = \frac{W_m}{W_m + W_d} \times 100 \quad \ldots\ldots\ldots\ (6.1)$$

Where, m = moisture content (%wb), W_m = weight of moisture in Kg and W_d = Weight of bone dry material in Kg.

Dry basis (db):

The % moisture content on dry basis is given by

$$M = \frac{W_m}{W_d} \times 100$$

$$\text{Or, } M = \left(\frac{m}{100 - m}\right) \times 100 \ldots\ldots\ldots\ (6.2)$$

Where, M = moisture content (%db)

Example 6.1: Drying of paddy

1. One thousand kilograms of parboiled paddy is to be dried from 32% to 13% moisture content (w.b.). Calculate the amount of moisture to be evaporated.

Sol^n:
Given,
Initial weight of paddy $(W_{m1} + W_d) = 1000$ kg
Initial moisture content of paddy in wet basis $(m_1) = 32\%$
Final moisture content of paddy in wet basis $(m_2) = 13\%$
Using equation 6.1, we can calculate water content at 32% (W_{m1}) as:

$$m_1 = \frac{W_{m1}}{W_{m1} + W_d} \times 100$$

$$\text{Or, } 32 = \frac{W_{m1} + W_d}{1000} \times 100$$

Or, $W_{m1} = 320$ kg.
Hence, the amount of bone –dry mass $(W_d) = 1000-320 = 680$ kg.
Again, using equation 6.1, we can calculate water content at 13% (W_{m2}) as:

$$m_2 = \frac{W_{m2}}{W_{m2} + W_d} \times 100$$

$$\text{Or, } 13 = \frac{W_{m2}}{W_{m2} + 680} \times 100$$

Or, $W_{m2} = 101.60$ kg
Hence, the amount of water evaporated $= W_{m1} - W_{m2} = 320 - 101.60 = 218.40$ kg.

Exercises on moisture content representations

1. 2 tonnes of paddy with 22% (wb) moisture content are to be dried to 13% (db) moisture content. Calculate the weight of bone – dry product and water evaporated [**Ans: 1560 Kg and 237.12 Kg**]

2. Determine the quantity of food material with 40% (wb) moisture content required to produce 1 tonne of product with 12% (wb) moisture content. [**Ans: 1466.608 kg**]

3. A fruit is dried from 85% to 40% moisture (wb). Express the moisture on dry basis and calculate the amount of water that is removed from 1000 kg of product. [**Ans: 566.66%, 66.66% & 750 kg**]

4. A sample of a food material weighing 20 kg is initially at 450% moisture content dry basis. It is dried to 25% moisture content wet basis. How much water is removed from the sample per kg of dry solids?

5. Calculate the amount of moisture evaporated from 100 kg of grain for drying it from an initial moisture content of 25% to a final moisture content of 13% on wet basis.

Theory of drying

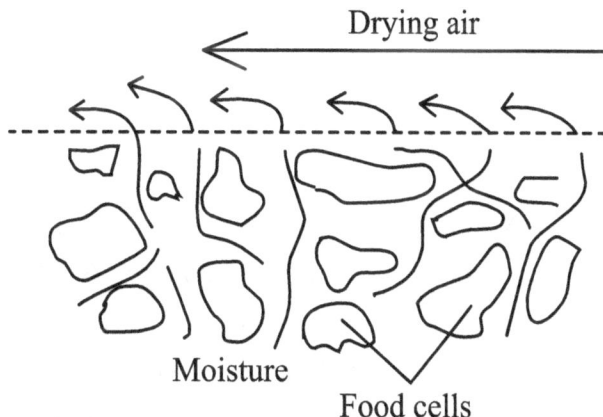

Fig. 6.2: Movement of moisture during drying

When hot air is blown over a wet food, heat is transferred to the surface, and latent heat of vapourization causes water to evaporate. Water vapour diffuses through a boundary film of air and is carried away by the moving air as shown in Fig. 6.2.

This creates a region of lower water vapour pressure at the surface of the food, and a water vapour pressure is established from the moist interior of the food to the dry air. This gradient provides the driving force for water removal from the food.

Water moves to the surface by the following mechanisms:
i. Liquid movement by capillary forces
ii. Diffusion of liquids, caused by differences in the concentration of solutes in different regions of the food
iii. Diffusion of liquid which is adsorbed in layers at the surface of solid components of the food, and
iv. Water vapour diffusion in air spaces within the food caused by vapour pressure gradients.

For a given food, the total amount of moisture that can be lost will vary with the humidity and temperature of the air. As water migrates out during drying, dissolved solids (sugar, acid, salt) are carried along to the surface. Here water evaporates into the air leaving the soluble solids which concentrate and may even precipitate at the surface.

As the drying proceeds, the water removal may be restrained by the drying process itself. Food tissue often sinks as it loses moisture and the structure may change and blocks the exit of water. Such a condition is known as case hardening in which the outer trough surface is formed but still moist interior remains. The hard outer surface is more impermeable to water and such a product is susceptible to microbial spoilage. Less intense drying and intermittent conditioning alleviate this problem.

Drying curve

During the drying of wet solid in heated air, the air supplies the necessary sensible and latent heat of evaporation of the moisture and also acts as a carrier gas for the removal of the water vapour formed from the vicinity of the evaporating surface.

Consider the situation where an inert solid, wetted with pure water, is being dried in a current of heated air flowing parallel to the drying surface. Assume that the temperature and humidity of the air above the drying surface remain constant throughout the drying cycle and that all the necessary heat is supplied to the material by convection. If the changes in moisture content of the material is recorded throughout drying, the data can be presented in the form of curves as given in Fig. 6.3. The study of these curves shows that the drying cycle can be considered in the following stages:

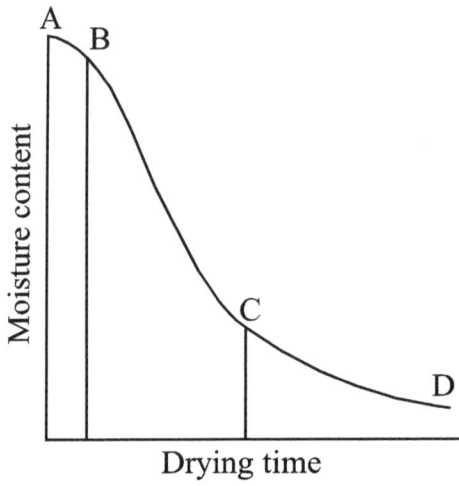

Fig. 6.3: Typical drying curve

A-B : This stage represents a settling down period during which the solid surface condition come into equilibrium with the drying air. It is often a negligible proportion of the overall drying cycle but in some cases it may be significant.

B-C : This is known as constant rate period and continues until a certain critical moisture content is reached. During this period, the surface of the solid remains saturated with liquid water by virtue of the fact that movement of water within the solid to the surface takes place at a rate as great as the rate of evaporation from the surface. Drying takes place by movement of water vapour from the saturated surface throughout a stagnant air film into the main stream of the drying air. The rate of drying is dependent on the rate of heat transfer to the drying surface. Here, the rate mass transfer balances the rate of heat transfer, and so the temperature of the drying surface remains constant. The three characteristics of air necessary for successful drying in the constant rate period are:

i. A moderately high DBT

ii. A low RH

iii. A high air velocity

C-D: As drying proceeds, a point is reached at which the rate of movement of moisture within the material to the surface is reduced to the extent that the surface begins to dry out. At this point, C, the rate of drying begins to fall and the falling rate period commences. From the point C onwards, the rate of drying slowly decreases until it approaches zero at the equilibrium moisture content (that is the food comes into equilibrium with the drying air). This is known as the falling rate period. During the falling rate period, the rate of water movement from the interior of the food to the surface falls below the rate at which water evaporates to the surrounding air. The surface therefore dries out. This is usually longest period of a drying operation and, in some foods (eg., grain drying) where the initial moisture content is below the critical moisture content, the falling rate period is the only part of the drying curve to be observed.

During the falling rate period the amount of water evaporating from the surface gradually decreases as heat is being supplied by the air, the surface temperature rises until it reaches the dry-bulb temperature of the drying air. Most heat damage to food therefore occurs in the falling rate period. A food dry until it reaches equilibrium moisture content to the surrounding air.

Drying rate curve

To obtain the rate of drying curve from drying curve, the slopes of the tangent drawn to the curve can be measured, which gives the value of dx/ dt at given values of 't'. The rate 'R' is calculated for each point by:

$$R = - \frac{L_s}{A} \cdot \frac{dx}{dt} \quad \ldots\ldots\ldots (6.3)$$

Where, R = drying rate in Kg water/ hm^2

L_s = Kg dry matter used

A = Exposed surface area for drying in m^2

Then, drying rate curve is obtained by plotting 'R' vs 'X'.

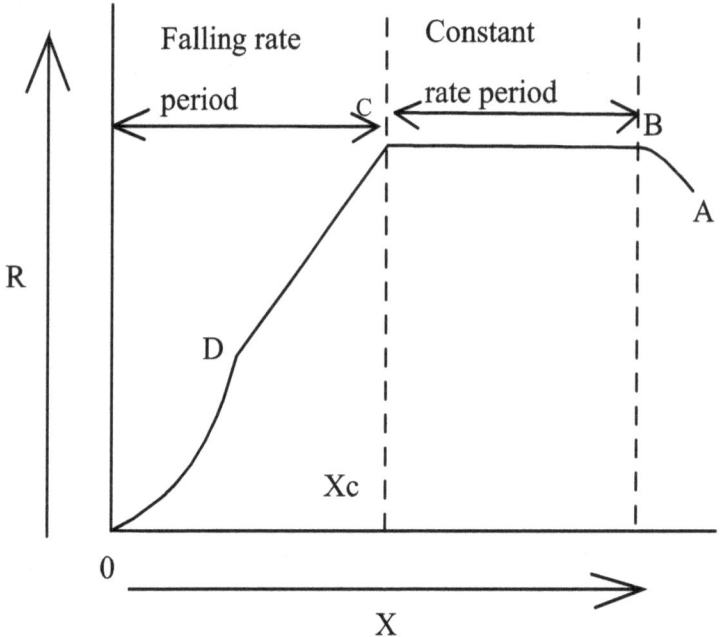

Fig. 6.4: Drying rate curve

Calculation of drying times:

a. <u>For constant rate period:</u>

From drying rate curve:

$$R = -\frac{L_s}{A} \cdot \frac{dx}{dt}$$

Or, $dt = \dfrac{L_s \cdot dx}{A \cdot R}$

Integrating the above equation:

$$\int_0^t dt = \frac{L_s}{A} \int_{X_c}^{X_1} \frac{dx}{R}$$

Or, $\boxed{t_c = \dfrac{L_s (X_1 - X_c)}{A \cdot R_c}}$ ………. (6.4)

Where R_c = Constant Drying rate in kgwater/ hr m²

b. <u>For falling rate period:</u>

$$\boxed{t_f = \frac{L_s}{A} \frac{X_c}{R_c} \ln \frac{X_c}{X_2}}$$ ………. (6.5)

Example 6.2: Drying of wet solid

1. A batch of wet solid is dried from a free m.c. of $X_1 = 0.38$ kg water/ kg dry solid to $X_2 = 0.04$ kg water/ kg dry solid. The weight of the dry solid is 399 kg and A = 18.56 m² of top drying surface. Calculate the time of drying if critical m.c. X_c is 0.195 kg water / kg dry solid and rate of drying under constant rate period R_c is 1.51 kg water/ hr m².

Sol^n:

Given,

Initial moisture content of solid (X_1) = 0.38 kg water/ kg dry solid
Final moisture content of solid (X_2) = 0.04 kg water/ kg dry solid
Weight of dry solid (L_s) = 399 kg
Area of drying surface (A) = 18.56 m²
Critical moisture content (X_c) = 0.195 kg water/ kg dry solid
Constant rate of drying (R_c) = 1.51 kg water/ hr m²
Total time of drying (t) = ?

Using equation 6.4, calculate constant rate drying time as:

$$t_c = \frac{L_s(X_1 - X_c)}{A \cdot R_c} = \frac{399(0.38 - 0.195)}{18.56 \times 1.51} = 2.63 \text{ hr}$$

Again, using equation 6.5, calculate falling rate drying time as:

$$t_f = \frac{L_s}{A} \frac{X_c}{R_c} \ln \frac{X_c}{X_2} = \frac{399 \times 0.195}{18.56 \times 1.51} \ln \frac{0.195}{0.04} = 4.06 \text{ hr}$$

Hence, $t = t_c + t_f = 6.69$ hr.

Hence, the total time required for drying is 6.69 hr.

Exercises on drying times

1. A material shows a constant drying rate of 0.15 kg water/ min kg dm and has a_w of 1 at m.c. above 1.10 kg water / kg dm. How long will it take to dry this material from an initial m.c. of 75% (wb) to a final m.c. of 8% (wb)? [**Ans:** 31.3 min]

2. Cut and blanched pieces of carrot are dehydrated in a cabinet dryer. The initial m.c of the carrot was 85% (wb) and it is to be dried to 5% m.c (wb). The critical moisture content is 30% (wb) and constant rate drying continues for 8 minutes. Estimate the total drying time for the product.

3. Calculate the time necessary to dry a product from 90% to 25% m.c (wb) in an industrial dryer where 2 kg dry solid/ m² surface area exposed to the air is loaded. It is given that critical m.c. is 5 kg water/ kg dry solid, EMC is 0.0333 kg water/ kg dry solid and the drying rate at the critical m.c is 3 kg water/ m²h under the specified drying conditions. [**Ans:** 11.96 hr]

4. Potato having 80% mc is dried to 15% mc and the CMC is 35%. 1000 Kg of potato was dried in a tray having an area of 12m² and the drying rate at constant period is 1.48. Calculate the time of drying.

5. The initial moisture content of a food product is 77% (wet basis), and the critical moisture content is 30% (wet basis). If the constant drying rate is 0.1 kg H_2O/(m²s), compute the time required for the product to begin the falling rate drying period. The product has a cube shape with 5 cm sides, and the initial product density is 950 kg/m³.[**Ans:** 53.2s]

6. A wet material having a critical moisture content of 15% and an equilibrium moisture content of 3% took 6 hours to dry from 45% to 5.5%. How long will it take to dry to 15%. All moisture contents are in dry basis.

7. A cabinet dryer is to be used for drying of a new food product. The product has an initial moisture content of 75% (wet basis) and requires 10 minutes to reduce the moisture content to a critical level of 30% (wet basis). Determine the final moisture of the product if a total drying time of 15 minutes is used.

Drying equipments
Tray dryer

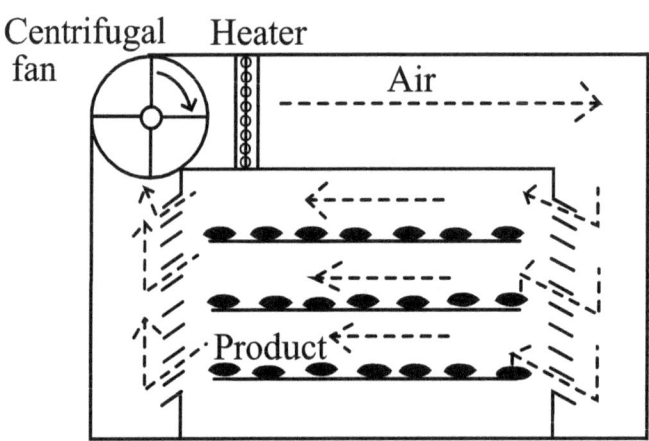

Fig. 6.5: Tray dryer

In tray dryers, the food is spread out, generally quite thinly, on trays in which the drying takes place as shown in Fig. 6.5. This consists essentially of an insulated cabinet containing an air circulating fan which moves the air through a heater and then through adjustable baffles which direct air either horizontally between the trays of food materials or vertically through the trays and food. Air heaters may be direct gas burners, steam coil exchangers or electrical resistance heaters. The air is blown past the heaters and thus heated air is used for dying. It is relatively cheap to build and maintain, flexible in design, and produces variable product quality due to relatively poor control. It is used singly or in groups, mainly for small-scale production (1-20 ton/day) of dried fruits and vegetables.

Tunnel dryers

These may be regarded as developments of the tray dryer, in which the trays on trolleys move through a tunnel where the heat is applied and the vapours removed. In most cases, air is used in tunnel drying and the material can move through the dryer either parallel or counter current to the air flow. Sometimes the dryers are compartmented, and cross-flow may also be used.

i. Concurrent system

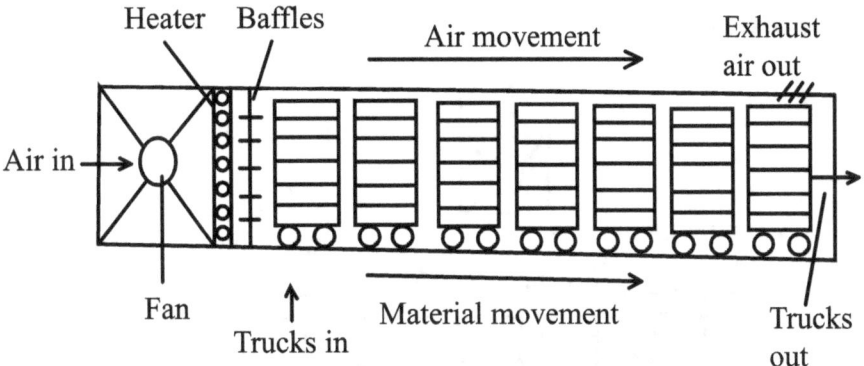

Fig. 6.6: Concurrent tunnel dryer

In this system, air direction is in the same direction of the material flow (parallel flow). This has the advantage that the hottest air contacts the wettest product, therefore high initial rate of drying is achieved at the wet end of the dryer which results in a product of low bulk density as little shrinkage occurs. On the other hand, the air at the dry end becomes cool and the moisture ladens, and the final product may not be sufficiently dry.

ii. Countercurrent system

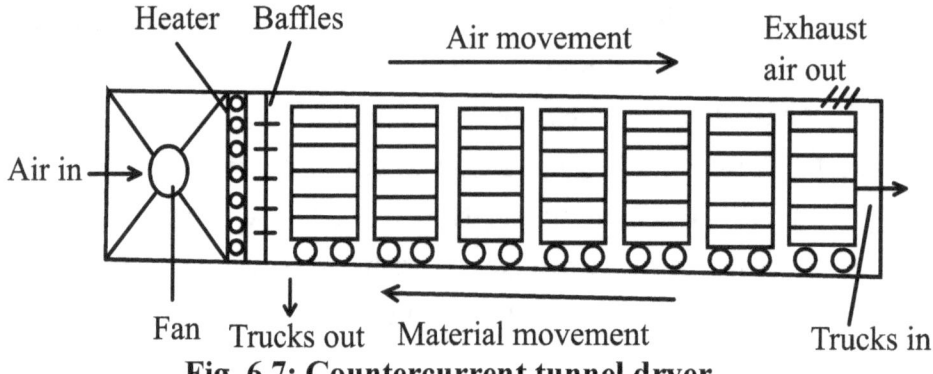

Fig. 6.7: Countercurrent tunnel dryer

In this system, air movement is in the opposite direction of the material flow. Low drying rates at the wet end results in high bulk density of the product. Overloading of the drier with the wet material can result in the long exposure of the food to a warm moist atmosphere and lead to spoilage. On the other hand, the dry product should not be left in the drier too long since it is in contact with the hottest air and could become overheated. This system is more economical in the use of heat than the concurrent system.

iii. Centre exhaust system

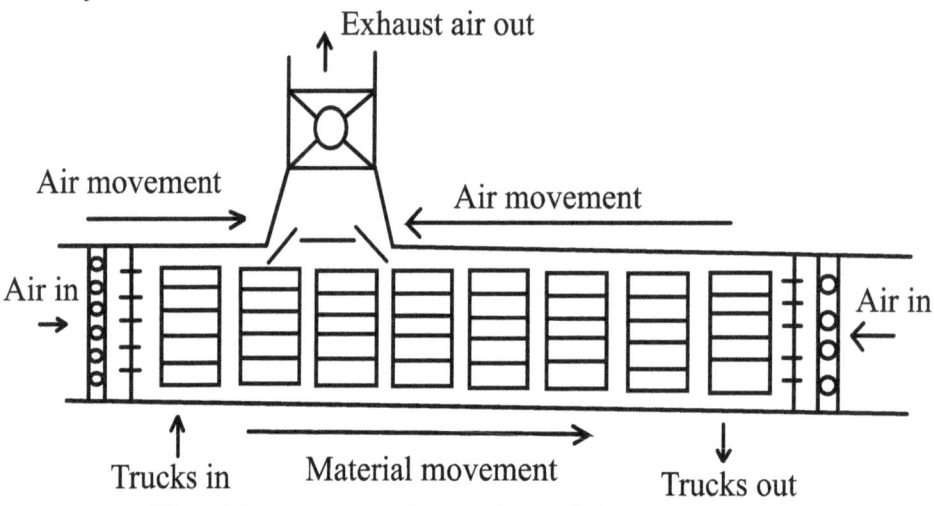

Fig. 6.8: Centre exhaust tunnel dryer

The previous two systems are combined into a single unit so as to derive their advantages. Shorter drying time, higher output, better control are attainable as compared with single stage units of smaller size, It incurs, however, higher capital cast as compared with single stage tunnels.

iv. Cross flow system

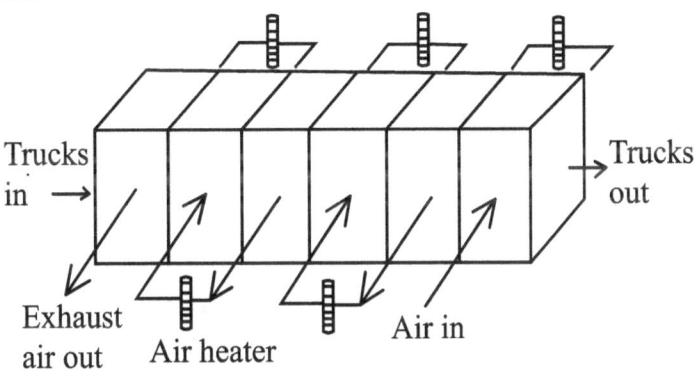

Exhaust air out   Air heater   Air in

Fig. 6.9: Cross flow tunnel dryer

In this system, the tunnel is divided into compartments, and air past the heater is passed through the belt loaded with material. Any one of the systems of air and material movements as in tunnel dryer may be used, but the most common one used in practice is the through flow system. Often upward movement is employed at the wet end of the tunnel, and downward air movement at the dry end to avoid lifting of the low density dry material.

Table 6.1: Advantages and limitations of parallel flow, counter-current flow, centre - exhaust and cross-flow drying

Type of air flow	Advantages	Limitations
Parallel or co-current type: Food ⟶ Air flow ⟶	Rapid initial drying. Little shrinkage of food. Low bulk density. Less heat damage to food. No risk of spoilage	Low moisture content difficult to achieve as cool moist air passes over dry food
Counter-current type: Food ⟶ Air flow ⟵	More economical use of energy. Low final moisture content as hot air passes over dry food	Food shrinkage and possible heat damage. Risk of spoilage from warm moist air meeting wet food
Cetre exhaust type: Food ⟶ Air flow ⟶↑⟵	Combined benefits of parallel and counter-current dryers but less than cross-flow dryers	More complex and expensive than single-direction air flow
Cross-flow type: Food ⟶ Air flow ↑↓	Flexible control of drying conditions by separately controlled heating zones, giving uniform drying and high drying rates	More complex and expensive to buy, operate and maintain

Roller or drum dryers

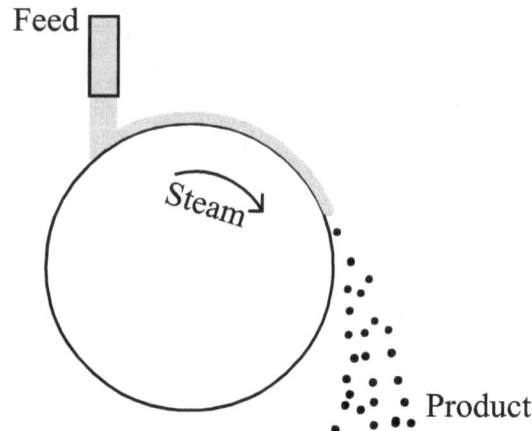

Fig. 6.10: Single stage drum dryer

In these dryer as shown in Fig. 6.10, the food is spread over the surface of a heated drum. The drum rotates, with the food being applied to the drum at one part of the cycle. The food remains on the drum surface for the greater part of the rotation, during which time the drying takes place, and is then scraped off. Drum drying may be regarded as conduction drying.

It consists of one or more internally heated (by steam or hot water) slowly rotating hollow steel drums on which a thin layer of food is spread uniformly over the outer surface by dipping, by spraying, by spreading or by auxillary feed rollers. Before the drum has completed 1 revolution,(within 20s-3min), the dried food is scrapped off by a doctor blade which contacts the drum surface uniformly along its length.

The single drum is widely used as it has greater flexibility, a larger proportion of the drum area available for drying, easier access for maintenance and no risk of damage caused by metal objects falling between the drums.

Drum dryers have high drying rates and high energy efficiencies. They are suitable for slurries in which the particles are too large for spray drying. However, the high capital cost of the machined drums, and heat damage to sensitive foods from high drum temperatures have caused a move to spray drying for many bulk dried foods.

Spray dryers

In a spray dryer as of Fig. 6.11, liquid or fine solid material in a slurry is sprayed in the form of a fine droplet dispersion into a current of heated air. Air and solids may move in parallel or counterflow. Drying occurs very rapidly, so that this process is very useful for materials that are damaged by exposure to heat for any appreciable length of time. The dryer body is large so that

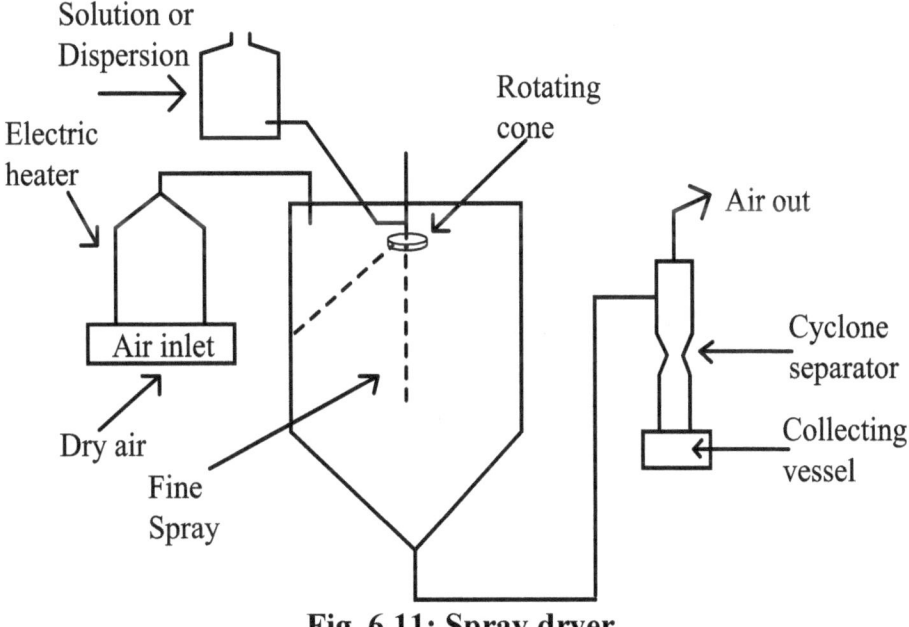

Fig. 6.11: Spray dryer

the particles can settle, as they dry, without touching the walls on which they might otherwise stick.

The principle of this dryer is to introduce a dispersion of pre concentrated food into drying chamber in the form of a fine spray (10-200μm in diameter) where it is brought into intimate contact with a stream of heater air (150-300°C), thereby achieving almost instantaneous drying (1-10 sec) and then to separate the dried product from air by cyclone separator.

The main advantages are rapid drying, large scale continuous production, low labour costs, and under the right conditions, it gives a very high quality product with little loss in nutritive properties.

The major limitations are high capital costs and only liquid or slurry type food can be dried. Eg., milk, egg, coffee, tea, cocoa, milk based baby foods, fruit and vegetable juices, edible proteins, meat and yeasts extract, wheat and corn products and other heat sensitive foods.

Fluidized bed dryers

The principle of this type of dryer is that heated air is forced up through a bed of particulate foods upto 0.15m deep under such conditions that the foods become suspended and vigorously agitated. The heated air thus acts as both the drying and the fluidizing medium, and thus maximum surface area of food is made available for drying.

If air is allowed to flow through a bed of solid powdered material in upward direction with the velocity greater than the settling rate of the particles, the solid particles will be blown up and become suspended in the air stream. At the stage, solid bed looks like the boiling liquid, therefore this stage is called as fluidised. Very high and mass transfer rates are obtained as a result of the intimate contact between individual particles and the fluidising gas.

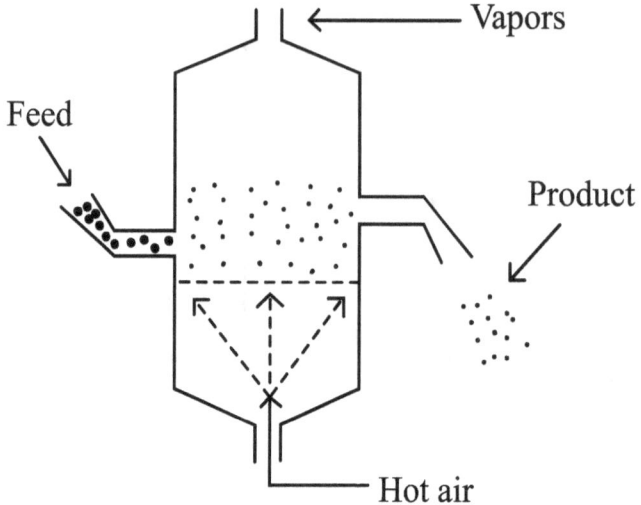

Fig. 6.12: Fluidized bed dryer

Fluidized bed dryers are compact, batch or continuous and have good control over drying conditions, relatively high thermal efficiencies and high drying rates. Fluidized bed dryers are limited to small particulate foods that are capable of being fluidized without excessive mechanical damage. E.g : peas, dice or sliced vegetables, grains, powders or extruded foods.

Chapter 7: Crystallization

Introduction

Crystallization is an example of a separation process in which mass is transferred from a liquid solution, whose composition is generally mixed, to a pure solid crystal. Soluble components are removed from solution by adjusting the conditions so that the solution becomes supersaturated and excess solute crystallizes out in a pure form. This is generally accomplished by lowering the temperature, or by concentration of the solution, in each case to form a supersaturated solution from which crystallization can occur. The equilibrium is established between the crystals and the surrounding solution, the mother liquor. The manufacture of sucrose, from sugar cane or sugar beet, is an important example of crystallization in food technology. Crystallization is also used in the manufacture of other sugars, such as glucose and lactose, in the manufacture of food additives, such as salt, and in the processing of foodstuffs, such as ice cream. In the manufacture of sucrose from cane, water is added and the sugar is pressed out from the residual cane as a solution. This solution is purified and then concentrated to allow the sucrose to crystallize out from the solution.

Magma

In industrial crystallization from solution, the two-phase mixture of mother liquor and crystals of all sizes, which occupies the cryatallizer and is withdrawn as product is called magma.

Crystallization equilibrium

Once crystallization is concluded, equilibrium is set up between the crystals of pure solute and the residual mother liquor, the balance being determined by the solubility (concentration) and the temperature. The driving force making the crystals grow is the concentration excess (supersaturation) of the solution above the equilibrium (saturation) level. The resistances to growth are the resistance to mass transfer within the solution and the energy needed at the crystal surface for incoming molecules to orient themselves to the crystal lattice.

Theory of crystallization

The overall process of crystallization from a supersaturated solution is considered to consist of the basic steps of nucleus formation or nucleation and of crystal growth.
If the solution is free of all solid particles, foreign or of the crystallizing substance, then nucleus formation must first occur before crystal growth starts. New nuclei may continue to form while the nuclei present are growing.

The driving force for the nucleation step and the growth step is super saturation. These two steps do not occur in a saturated or unsaturated solution.

Nucleation theory

Primary nucleation is a result of rapid local fluctuations on a molecular scale in a homogenous phase.

Crystal nuclei may form from various kinds of particles: molecules, atoms or ions. In aqueous solutions these may associate to form what is called a cluster- a rather loose aggregation that usually disappears quickly. Occasionally, however, enough particles associate into what is known as an embryo, in which there are the beginnings of a lattice arrangement and the formation of a new and separate phase.

If the supersaturation is large enough, an embryo may grow to such a size that is in thermodynamic equilibrium with the solution. It is then called a nucleus which is the smallest assemblage of particles that will not redissolve and can therefore grow to form a crystal.

Cluster ⟶ Embryo ⟶ Nucleus ⟶ Crystal

Miers theory explain the formation of nuclei and crystals in an unseeded solution as given in Fig. 7.1.

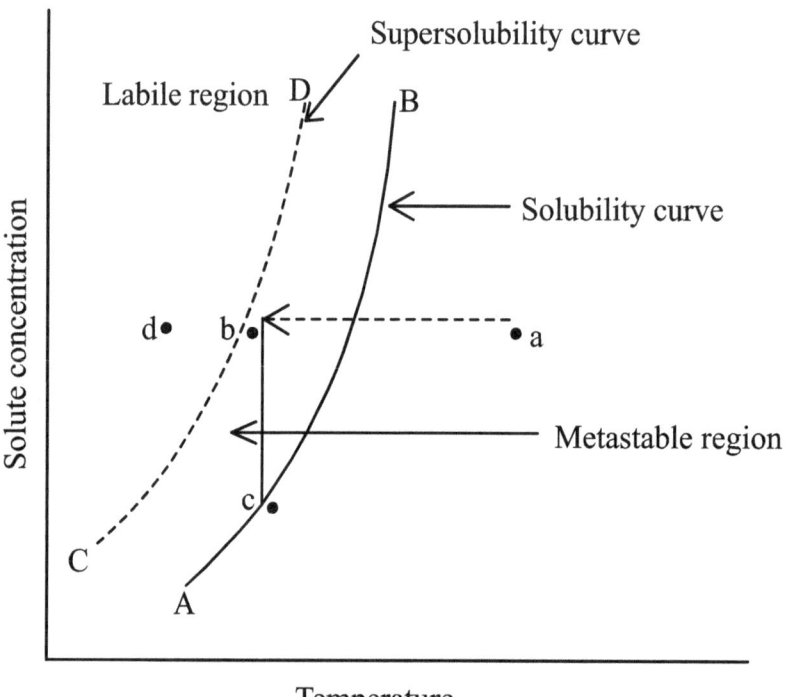

Fig. 7.1: Mier's theory of crystallization

Line AB is the normal solubility curve. If a sample of solution at point a is cooled, it first crosses the solubility curve. The sample will not crystallize until it has supercooled to some point b where crystallization begins, and the concentration drops to point c if no further cooling is done. The curve CD, called the supersolubility curve represents the limit at which nucleus formation starts spontaneously and hence where crystallization can start. Any crystal in the metastable region will grow.

Secondary or contact nucleation, which is the most effective method of nucleation occurs when crystal collide with each other, with the impellers in mixing, or with the walls of the pipe or containers. This type of nucleation is, of course, affected by the intensity of agitation. It occurs at low super saturation, where the crystal growth rate is at the optimum for good crystals.

Rate of crystal growth

Once nucleii are formed, either spontaneously or by seeding, the crystals will continue to grow so long as supersaturation persists. The three main factors controlling the rates of both nucleation and of crystal growth are the temperature, the degree of supersaturation and the interfacial tension between the solute and the solvent. If supersaturation is maintained at a low level, nucleus formation is not encouraged but the available nucleii will continue to grow and large crystals will result. If supersaturation is high, there may be further nucleation and so the growth of existing crystals will not be so great. In practice, slow cooling maintaining a low level of supersaturation produces large crystals and fast cooling produces small crystals.

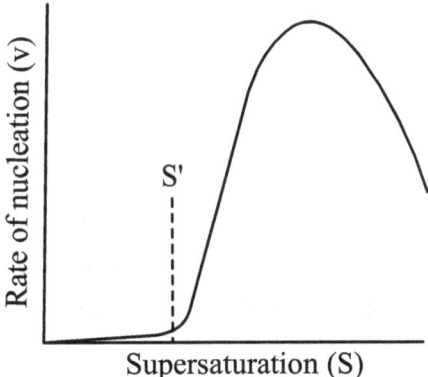

Fig. 7.2: Rate of crystal growth

Nucleation rate is also increased by agitation. For example, in the preparation of fondant for cake decoration, the solution is cooled and stirred energetically. This causes fast formation of nucleii and a large crop of small crystals, which give the smooth texture and the opaque appearance desired by the cake decorator.

Once nucleii have been formed, the important fact in crystallization is the rate at which the crystals will grow. This rate is controlled by the diffusion of the solute through the solvent to the surface of the crystal and by the rate of the reaction at the crystal face when the solute molecules rearrange themselves into the crystal lattice.

These rates of crystal growth can be represented by the equations

$$\frac{dw}{dt} = K_d A(c - c_i) \quad\quad\quad (7.1)$$

$$\frac{dw}{dt} = K_s(c_i - c_s) \quad\quad\quad (7.2)$$

where dw is the increase in weight of crystals in time dt, A is the surface area of the crystals, c is the solute concentration of the bulk solution, c_i is the solute concentration at the crystal/solution interface, c_s is the concentration of the saturated solution, K_d is the mass transfer coefficient to the interface and K_s is the rate constant for the surface reaction.

These equations are not easy to apply in practice because the parameters in the equations cannot be determined and so the equations are usually combined to give:

$$\frac{dw}{dt} = KA(c - c_s) \quad\quad\quad (7.3)$$

Where, $\dfrac{1}{K} = \dfrac{1}{K_d} + \dfrac{1}{K_s}$

or

$$\boxed{\frac{dL}{dt} = \frac{K(c - c_s)}{\rho_s}} \quad\quad\quad (7.4)$$

since, $dw = A\rho_s dL$

and dL/dt is the rate of growth of the side of the crystal and ρ_s is the density of the crystal.

It has been shown that at low temperatures diffusion through the solution to the crystal surface requires only a small part of the total energy needed for crystal growth and, therefore, that diffusion at these temperatures has relatively little effect on the growth rate. At higher temperatures, diffusion energies are of the same order as growth energies, so that diffusion becomes much more important. Experimental results have shown that for sucrose the limiting temperature is about 45°C, above which diffusion becomes the controlling factor. Impurities in the solution retard crystal growth; if the concentration of impurities is high enough, crystals will not grow.

L- law of crystal growth

Batch crystallizers often are seeded with small crystals of a known range of sizes. The resulting CSD for a given overall weight gain can be estimated by an approximate relation known as the McCabe Delta-L Law, which states that each original crystal grows by the same amount ΔL:

a. All crystals have the same shape.
b. They grow invariantly, i.e. the growth rate is independent of crystal size.
c. Supersaturation is constant throughout the crystallizer.
d. No nucleation occurs.
e. No size classification occurs in the crystallizer
f. The relative velocity between crystals and liquor remains constant.

The relation between the relative masses of the original and final size distributions is given in terms of the incremental ΔL by

$$R = \frac{\sum w_i \left(L_{0i} + \Delta L\right)^3}{\sum w_i L_{0i}^3} \quad \ldots\ldots\ldots (7.5)$$

where R: ratio of final and initial weight of crystal, w_i: fraction of crystal of size L_i
L_{0i}: initial dimension of crystal i , L_i : final dimension of crystal i

When R is specified, ΔL is found by trial, and then the size distribution is evaluated.

Solubility and saturation

Solubility is defined as the maximum weight of anhydrous solute that will dissolve in 100 g of solvent. In the food industry, the solvent is generally water.

Solubility is a function of temperature. For most food materials increase in temperature increases the solubility of the solute as shown for sucrose in Fig. 7.3. Pressure has very little effect on solubility.

During crystallization, the crystals are grown from solutions with concentrations higher than the saturation level in the solubility curves. Above the supersaturation line, crystals form spontaneously and rapidly, without external initiating action. This is called spontaneous nucleation. In the area of concentrations between the saturation and the supersaturation curves, the metastable region, the rate of initiation of crystallization is slow; aggregates of molecules form but then disperse again and they will not grow unless seed crystals are added. Seed crystals are small crystals, generally of the solute, which then grow by deposition on them of further solute from the

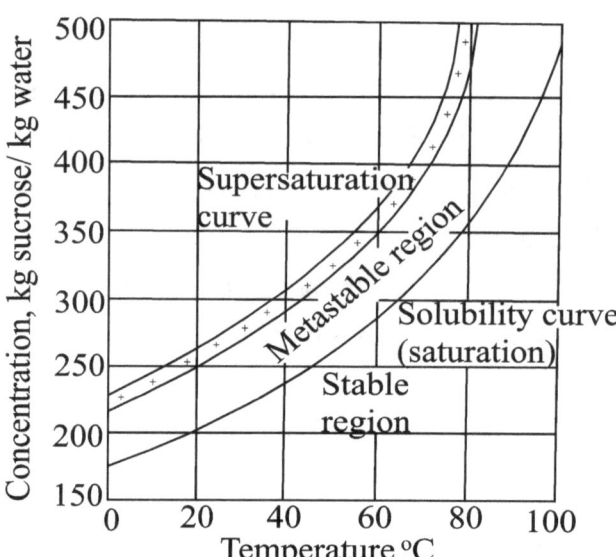

Fig. 7.3: Solubility and saturation curves for sucrose in water

solution. This growth continues until the solution concentration falls to the saturation line. Below the saturation curve there is no crystal growth, crystals instead dissolve.

Example 7.1: Solubility of sodium chloride

1. If sodium chloride solution, at a temperature of 40°C, has a concentration of 50% when the solubility of sodium chloride at this temperature is 36.6 g / 100 g water, calculate the quantity of sodium chloride crystals that will form once crystallization has been started.

Solution: Weight of salt in solution = 50 g / 100 g solution
= 100 g / 100 g water.
Saturation concentration = 36.6 g / 100 g water
Weight crystallized out = (100 - 36.6) g / 100 g water = 63.4 g / 100 g water

To remove more salt, this solution would have to be concentrated by removal of water, or else cooled to a lower temperature.

Mass and heat balance

Most crystallization processes are carried out slowly, and the surface of solid material in contact with the solution is so great that at the end of the process, the mother liquor is saturated at the terminal temperature. It is true for some materials that crystallizes very slowly, such as sucrose, the solution can support a considerable supersaturation even when in contact with solid sugar for an appreciable length of time. In general, however, the final concentration of the mother liquor can be taken as that read from the solubility curve. The yield of a crystallization process can therefore be calculated from the solubility data, if the initial concentration and the final temperature of the solution are known.

Calculation of the yield in the case of a substance that comes out in the anhydrous form is simple. It is necessary only to take the difference between the initial composition of the solution and the solubility corresponding to the final temperature to get the yield. Both concentrations must be expressed in terms of the water content of the solution and not in percent solid because the water content of the solution is the inert material that goes through the process unchanged.

For material balance, express all compositions in terms of hydrated salt and excess water, since it is the latter quantity that remains constant during the crystallization process and compositions expressed on the basis of this excess water content are substracted to give a correct result.

Example 7.2: Crystallization of salt

1. A salt solution weighing 10000kg with 30 wt% Na_2CO_3 is cooled to 293K. The salt crystallizes as decahydrate. What will be the yield of Na_2CO_3 crystals if solubility is 21.5 kg anhydrous Na_2CO_3/100 Kg of total water? Do this for the following cases:

a. Assume that no water is evaporated.

b. Assume that 3% of total weight of solution is lost by evaporation of water in cooling?

Sol^n:

Given,

Weight of salt soln (F) =10000 kg

Conc^n of Na_2CO_3 in feed (x_F) =30%

Molecular weight of Na_2CO_3 =106

Molecular weight of $Na_2CO_3.10H_2O$ =286.2

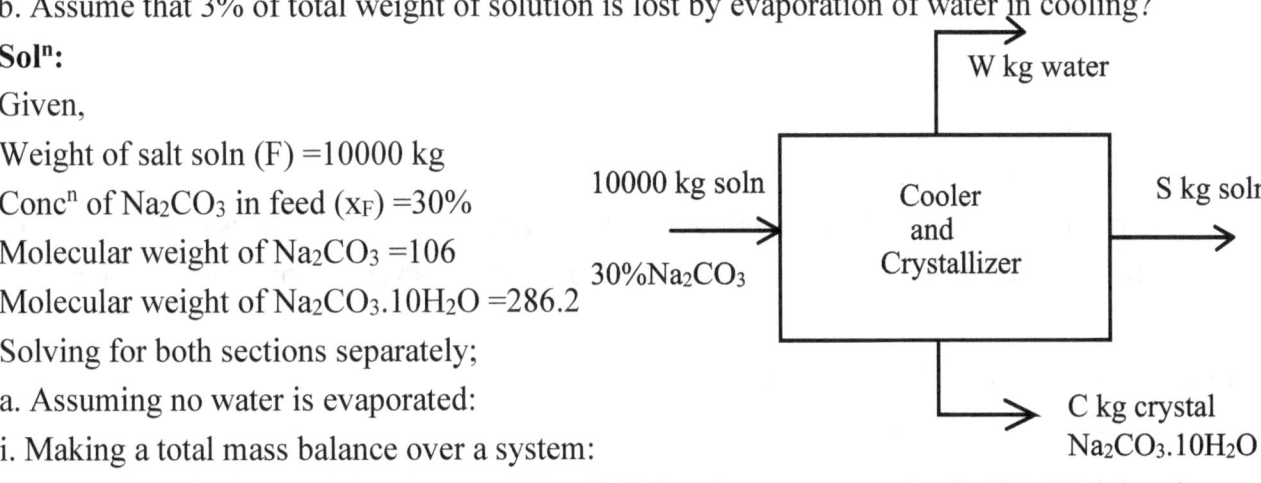

Solving for both sections separately;

a. Assuming no water is evaporated:

i. Making a total mass balance over a system:

Weight of Feed (F) = Weight of water (W) + Weight of supersaturated sol^n (S) + Weight of crystal (C)

Or, F= W +S +C

Or, 10000 = 0 + S +C

Or, S + C =10000.......... (7.6)

ii. Making a component mass balance (of Na_2CO_3):

135

$$10000 \times 0.3 = S \times \left(\frac{21.5}{100+21.5}\right) + C \times \left(\frac{106}{286.2}\right)$$

Or, $3000 = S \times \left(\frac{21.5}{121.5}\right) + C \times \left(\frac{106}{286.2}\right)$ .... **(7.7)**

Solving equation (7.6) and (7.7):

S = 3636 kg and C = 6364.46 kg.

b. Assuming that 3% of total weight of solution is lost by evaporation of water in cooling:

Weight of water (W) = 3% of 10000 = 300 kg

i. Making a total mass balance over a system:

Weight of Feed (F) = Weight of water (W) + Weight of supersaturated sol^n (S) + Weight of crystal (C)

Or, F = W + S + C

Or, 10000 = 300 + S + C

Or, S + C = 9700.......... **(7.8)**

ii. Making a component mass balance (of Na_2CO_3):

$$10000 \times 0.3 = W \times 0 + S \times \left(\frac{21.5}{100+21.5}\right) + C \times \left(\frac{106}{286.2}\right)$$

Or, $3000 = S \times \left(\frac{21.5}{121.5}\right) + C \times \left(\frac{106}{286.2}\right)$ .... **(7.9)**

Solving equation (7.8) and (7.9):

S = 3070 kg and C = 6630 kg.

Hence, the yield of crystal is found to be 6364.46 kg assuming no water is evaporated, and 6630 kg when we assume 3% of total weight of solution is lost by evaporation of water in cooling.

Exercises on mass and heat balance in crystallizer

1. What will be the yield of hypo ($Na_2S_2O_3.5H_2O$) if 100 lb of a 48 percent solution of $Na_2S_2O_3$ are cooled to 68°F. The solubility at 68°F is 70 parts $Na_2S_2O_3$ per 100 parts of water. [**Ans:** 30.2lb]

2. A solution of $MgSO_4$ at 220°F containing 43 g $MgSO_4$ per 100 g H_2O is fed into a cooling crystallizer operating at 50°F. If the solution leaving the crystallizer is saturated, what is the rate at which the solution must be fed to the crystallizer to produce one ton of $MgSO_4 \cdot 7H_2O$ per hour? The solubility at 50°F is 30 parts $MgSO_4$ per 100 parts of water. [**Ans:** 3.65 ton/h]

3. 1,200 lb of barium nitrate are dissolved in sufficient water to form a saturated solution at 90°C. Assuming that 5% of the weight of the original solution is lost through evaporation, calculate the crop of the crystals obtained when cooled to 20°C. Solubility data of barium nitrate at 90°C = 30.6 lb/100 lb water; at 20°C = 9.2 lb/100 lb water. [**Ans:** 862.9894 lb]

4. A hot solution of $Ba(NO_3)_2$ from an evaporator contains 30.6 kg $Ba(NO_3)_2$/100 kg H_2O and goes to a crystallizer where the solution is cooled and $Ba(NO_3)_2$ crystallizes. On cooling, 10% of the original water present evaporates. For a feed solution of 100 kg total, calculate the following:
a. The yield of crystals if the solution is cooled to 290K, where the solubility is 8.6 kg $Ba(NO_3)_2$/100 kg total water
b. The yield if cooled instead to 283K, where the solubility is 7 kg $Ba(NO_3)_2$/100 kg total water
[**Ans:** 17.5026 kg & 18.6077 kg]

5. A batch of 1,000 kg of KCl is dissolved in sufficient water to make a saturated solution at 363 K, where the solubility is 35 wt % KCl in water. The solution is cooled to 293 K, at which temperature its solubility is 25.4 wt %.
a. What are the weight of water required for the solution and the weight of KCl crystals obtained?
b. What is the weight of crystals obtained if 5% of the original water evaporates on cooling? [**Ans:** C=367.67 kg & wt H_2O=1,857.14 kg; C=399.29 kg]

6. Crystals of $Na_2CO_3 \cdot 10H_2O$ are dropped into a saturated solution of Na_2CO_3 in water at 100°C. What percent of the Na_2CO_3 in the $Na_2CO_3 \cdot H_2O$ is recovered in the precipitated solid? The precipitated solid is $Na_2CO_3 \cdot H_2O$. Data at 100°C: the saturated solution is 31.2% Na_2CO_3; molecular weight of Na_2CO_3 is 106. [**Ans:** 22.99%]

7. The solubility of sodium bicarbonate in water is 9.6 g per 100 g water at 20°C and 16.4 g per 100 g water at 60°C. If a saturated solution of $NaHCO_3$ at 60°C is cooled to 20°C, what is the percentage of the dissolved salt that crystallizes out? [**Ans:** 41.46%]

8. A feed solution of 2268 kg at 327.6 K (54.4°C) containing 48.2 kg $MgSO_4$/100 kg total water is cooled to 293.2 K (20°C), where $MgSO_4 \cdot 7H_2O$ crystals are removed. The solubility of the salt is 35.5 kg $MgSO_4$/100 kg total water. The average heat capacity of the feed solution can be assumed as 2.93 kJ. The heat of solution at 291.2 K (18°C) is 13.31×10^3 kJ (kg mol $MgSO_4 \cdot 7H_2O$. Calculate the yield of crystals and make a heat balance to determine the total heat absorbed,

q, assuming that no water is vapourized. [**Ans:** C = 6 1 6.9 kg $MgSO_4 \cdot 7H_2O$ crystals and S = 1 6 5 1.1 kg solution; - 26 1 9 1 2 kJ]

9. Lactose syrup is concentrated to 8 g lactose per 10 g of water and then run into a crystallizing vat which contains 2,500 kg of the syrup. In this vat, containing 2,500 kg of syrup, it is cooled from 57°C to 10°C. Lactose crystallizes with one molecule of water of crystallization. The specific heat of the lactose solution is 3470 J/kg°C. The heat of solution for lactose monohydrate is -15,500 kJ/kmol. The molecular weight of lactose monohydrate is 360 and the solubility of lactose at 10°C is 1.5 g/10 g water. Assume that 1% of the water evaporates and that the heat loss trough the vat walls is 4 x 10^4 kJ. Calculate the heat to be removed in the cooling process. [**Ans:** 454.72 x 10^3 kJ]

Crystallization processes in the food industries

Classification of crystallizers: Crystallization equipment is most easily classified by the methods by which supersaturation is brought about. These are as follows:

Supersaturation by cooling

It can be used for those substances that have a solubility curve that decreases appreciably with temperature. If the solubility of the solute increases strongly with increase in temperature, as in the case with many common inorganic salts and organic substances, a saturated solution becomes supersaturated by simple cooling and temperature reduction.

Supersaturation by the evaporation of the solvent

Superaturation by the evaporation of the solvent find its principal application in the production of common salt, where the solubility curve is so flat the yield of solids by cooling would be negligible.

Supersaturation by adiabatic evaporation (Cooling plus evaporation)

The third method, namely cooling adiabatically in a vacuum is the most important method for large- scale production. If a hot solution is introduced into a vacuum where the total pressure is less than the vapour pressure of the solvent at the temperature of which it is introduced, the solvent must flash; and the flashing must produce adiabatic cooling. The combination of evaporation and cooling produces the desired supersaturation.

Salting out technique

The last method may act physically by forming, with the original solvent, a mixed solvent in which the solubility of the solvent is sharply reduced. This process is called salting. Or, if a nearly complete precipitation is required, a new solute may be created chemically by adding a third component that will react with the original solute to form an insoluble substance. This process is called precipitation.

Classification of crystallizers

According to the operations in which the solid is the desired material, crystallizers can be classified into the following described types:

Tank crystallizers

For many years the common practice in producing crystals was to prepare hot, nearly saturated solutions and run these solutions into open rectangular tanks in which the solution stood while it cooled and deposited crystals. No attempt was usually made to seed these tanks, to provide for agitation, or to accelerate or control the crystallization in any way.

Sometimes rods or strings were hung in the tanks to give the crystals additional surface on which to grow and to keep at least a part of the product out of the sediment that might collect in the bottom of the tank.

When the tank had cooled sufficiently which was usually a matter of several days, any remaining mother liquor was drained off and the crystals removed by hand. This involved much labor and often resulted in the inclusion, with the crystals, of any impurities that settles to the bottom of the tank. The floor space required and the amount of material tied up in the process were both large.

Agitated batch crystallizers

The old method of growing crystals was wasteful of material, labor, and floor space, and artificial cooling was desirable. Water is circulated through the cooling coils, and the solution is agitated by the propellers on the central shaft. This agitation performs two functions:

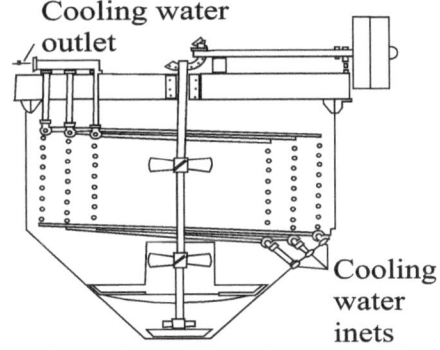

Fig. 7.4: Agitated batch crystallizer

Firstly, it increases the rate of heat transfer and keeps the temperature of the solution more nearly uniform and secondly, by keeping the fine crystals in suspension it gives them an opportunity to grow uniformly instead of forming large crystals or aggregates.

Vacuum crystallizers

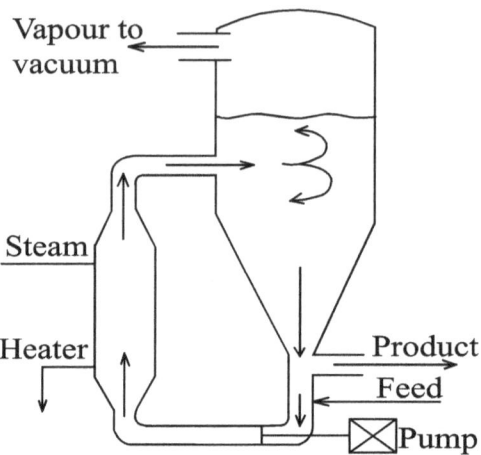

Fig. 7.5: Vacuum crystallizer

If a warm saturated solution be introduced into a vessel in which a vacuum is maintained that corresponds to a boiling point of the solution lower than feed temperature, the solution so introduced must flash and be cooled by the resulting adiabatic evaporation. Not only will the resultant cooling causes crystallization, but also there is some evaporation taking place at the same time which thereby increases the yield.

Chapter 8: Filtration

Introduction

In another class of mechanical separations, placing a screen in the flow through which they cannot pass imposes virtually total restraint on the particles above a given size. The fluid in this case is subject to a force that moves it past the retained particles. This is called filtration. The particles suspended in the fluid, which will not pass through the apertures, are retained and build up into what is called a filter cake. Sometimes it is the fluid, the filtrate that is the product, in other cases the filter cake.

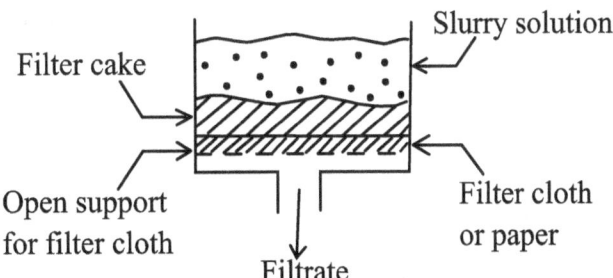

Fig. 8.1: Simple laboratory filtration apparatus

The fine apertures necessary for filtration are provided by fabric filter cloths, by meshes and screens of plastics or metals, or by beds of solid particles. In some cases, a thin preliminary coat of cake, or of other fine particles, is put on the cloth prior to the main filtration process. This preliminary coating is put on in order to have sufficiently fine pores on the filter and it is known as a pre-coat.

Basic theory of filtration

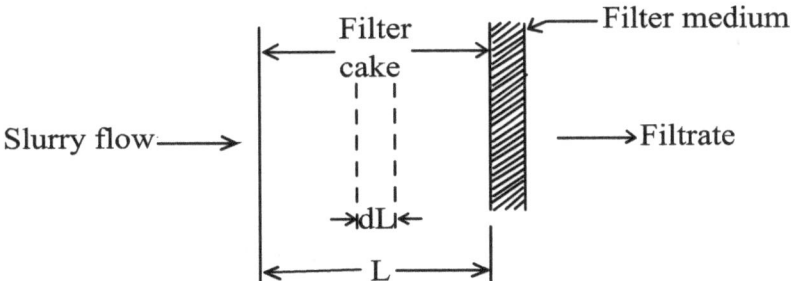

Fig. 8.2: Basic theory of filtration

The analysis of filtration is largely a question of studying the flow system. The fluid passes through the filter medium, which offers resistance to its passage, under the influence of a force which is the pressure differential across the filter. Thus, we can write the familiar equation:

$$\text{Rate of filtration} = \frac{\text{driving force}}{\text{resistance}}$$

Resistance arises from the filter cloth, mesh, or bed, and to this is added the resistance of the filter cake as it accumulates. The filter-cake resistance is obtained by multiplying the specific resistance of the filter cake that is its resistance per unit thickness, by the thickness of the cake. The resistances of the filter material and pre-coat are combined into a single resistance called the filter resistance. It is convenient to express the filter resistance in terms of a fictitious thickness of filter cake. This thickness is multiplied by the specific resistance of the filter cake to give the filter resistance. Thus the overall equation giving the volumetric rate of flow dV/dt is:

$$\frac{dV}{dt} = \frac{A\Delta P}{R} \quad \ldots\ldots\ldots (8.1)$$

As the total resistance is proportional to the viscosity of the fluid, we can write:

$$R = \mu r (L_c + L) \quad \ldots\ldots\ldots (8.2)$$

where R is the resistance to flow through the filter, μ is the viscosity of the fluid, r is the specific resistance of the filter cake, L_c is the thickness of the filter cake and L is the fictitious equivalent thickness of the filter cloth and pre-coat, A is the filter area, and ΔP is the pressure drop across the filter.

If the rate of flow of the liquid and its solid content are known and assuming that all solids are retained on the filter, the thickness of the filter cake can be expressed by:

$$L_c = \frac{wV}{A} \quad \ldots\ldots\ldots (8.3)$$

where, w is the fractional solid content per unit volume of liquid, V is the volume of fluid that has passed through the filter and A is the area of filter surface on which the cake forms.

The resistance can then be written as using equation 8.2 and 8.3;

$$R = \mu r \left\{ w\left(\frac{V}{A}\right) + L \right\} \quad \ldots\ldots\ldots (8.4)$$

and the equation for flow through the filter, under the driving force of the pressure drop is then using equation 8.1 and 8.4;

$$\frac{dV}{dt} = \frac{A\Delta P}{\mu r \left\{ w\left(\frac{V}{A}\right) + L \right\}} \quad \ldots\ldots\ldots (8.5)$$

Equation 8.5 may be regarded as the fundamental equation for filtration. It expresses the rate of filtration in terms of quantities that can be measured, found from tables, or in some cases estimated. It can be used to predict the performance of large-scale filters on the basis of laboratory or pilot scale tests. Two applications of equation 8.5 are filtration at a constant flow rate and filtration under constant pressure.

Constant-rate filtration

In the early stages of a filtration cycle, it frequently happens that the filter resistance is large relative to the resistance of the filter cake because the cake is thin. Under these circumstances, the resistance offered to the flow is virtually constant and so filtration proceeds at a more or less constant rate. Equation 8.5 can then be integrated to give the quantity of liquid passed through the filter in a given time. The terms on the right-hand side of equation 8.5 are constant so that integration is very simple:

$$\int \frac{dV}{Adt} = \frac{V}{At} = \frac{\Delta P}{\mu r \left\{ w \left(\frac{V}{A} \right) + L \right\}}$$

$$\text{Or,} \quad \boxed{\Delta P = \frac{V}{At} \times \mu r \left\{ w \left(\frac{V}{A} \right) + L \right\}} \quad \ldots\ldots\ldots (8.6)$$

From equation 8.6, the pressure drop required for any desired flow rate can be found. Also, if a series of runs is carried out under different pressures, the results can be used to determine the resistance of the filter cake.

If 'L' is assumed to be of negligible thickness, equation 8.6 can be written as:

$$\Delta P = \frac{\mu r w V^2}{A^2 t} \quad \ldots\ldots\ldots (8.7)$$

Example 8.1: Constant rate filtration in an air filter

1. An air filter is used to remove small particles from an air supply to a quality control laboratory. The air is supplied at the rate of 0.5 m³/s through an air filter with a 0.5 m² cross-section. If the pressure drop across the filter is 0.25 cm water after 1 hr of usage, determine the life of filter if a filter change is required when the pressure drop is 2.5 cm of water.

Sol^n:
Given,
Volumetric flow rate of air (Q) = 0.5 m³/s
Area of air- filter (A) = 0.5 m²
Pressure drop across the filter (ΔP) = 0.25 cm of water = 24.5 Pa
Time of usage (t) = 1 hr = 3600s
∴ Volume of filtrate (V) = Q × t = 0.5 × 3600 = 1800 m³
Viscosity of the fluid (μ), specific resistance of the filter cake (r) and the fractional solid content per unit volume of liquid (w) remains constant for the same materials being used.
For constant-rate filtration, using equation 8.7;

$$\Delta P = \frac{\mu r w V^2}{A^2 t}$$

Or, $\mu r w = \dfrac{A^2 t \Delta P}{V^2} = \dfrac{(0.5)^2 \times 3600 \times 24.5}{(1800)^2} = 0.006805 \text{ Pa.s.m}^{-2}$

Again,
New pressure drop across a filter (ΔP$_n$) = 2.5 cm of water = 245 Pa
Time of usage (t$_n$) = ?
∴ Volume of filtrate (V$_n$) = Q × t$_n$
Again using equation 8.7, we have:

$$\Delta P_n = \frac{\mu r w V_n^2}{A^2 t_n} = \frac{\mu r w (Q \times t_n)^2}{A^2 t_n}$$

Or, $t_n = \dfrac{\Delta P_n A^2}{\mu r w Q^2} = \dfrac{245 \times (0.5)^2}{0.006805 \times (0.5)^2} = 36002.9390 \text{ s} = 10 \text{ hrs}$

Hence, the filter change is required after 10 hrs of usage.

Constant-pressure filtration

Once the initial cake has been built up, and this is true of the greater part of many practical filtration operations, flow occurs under a constant-pressure differential. Under these conditions, the term ΔP in equation. 8.5 is constant and so;

$$\mu r \left\{ w\left(\frac{V}{A}\right) + L \right\} dV = A \Delta P dt$$

and integration from V = 0 at t = 0, to V = V at t = t

$$\mu r\left\{w\left(\frac{V^2}{2A}\right)+LV\right\}=A\Delta Pt \text{ and rewriting this}$$

$$\frac{tA}{V}=\frac{\mu rw}{2\Delta P}\times\frac{V}{A}+\frac{\mu rL}{\Delta P}$$

Or, $$\boxed{\frac{t}{\left(\frac{V}{A}\right)}=\frac{\mu rw}{2\Delta P}\times\frac{V}{A}+\frac{\mu rL}{\Delta P}}\ \ldots\ldots\ldots (8.8)$$

If 'L' is assumed to be of negligible thickness, equation 8.8 can be written as:

$$\Delta P=\frac{\mu rwV^2}{2A^2t}\ \ldots\ldots\ldots (8.9)$$

Equation 8.8 is useful because it covers a situation that is frequently found in a practical filtration plant. It can be used to predict the performance of filtration plant on the basis of experimental results. If a test is carried out using constant pressure, collecting and measuring the filtrate at measured time intervals, a filtration graph can be plotted of t/(V/A) against V/A and from the statement of equation 8.8, it can be seen that this graph should be a straight line. The slope of this line will correspond to $\mu rw/2\Delta P$ and the intercept on the t/(V/A) axis will give the value of $\mu rL/\Delta P$. Since, in general, μ, w, ΔP and A are known or can be measured, the values of the slope and intercept on this graph enable L and r to be calculated.

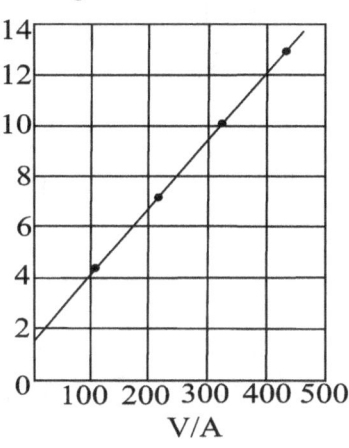

Example 8.2: Constant pressure filtration in a filter

1. A liquid is filtered at a pressure of 200KPa through a 0.2 m² filter. Initial results indicate that 5 mins is required to filter 0.3 m³ of liquid. Determine the time that will elapse until the rate of filtration drops to 5×10^{-5} m³/s.

Solⁿ:

Given,

Constant pressure drop across the filter (ΔP) = 200KPa = 200000 Pa

Area of filter (A) = 0.2 m²

Time of filtration (t) = 5 mins = 300s

Volume of filtrate (V) = 0.3 m³

Viscosity of the fluid (μ), specific resistance of the filter cake (r) and the fractional solid content per unit volume of liquid (w) remains constant for the same materials being used.

For constant-pressure filtration, using equation 8.9;

$$\Delta P = \frac{\mu r w V^2}{2A^2 t}$$

Or, $\mu r w = \frac{2\Delta P A^2 t}{V^2} = \frac{2 \times 200000 \times (0.2)^2 \times 300}{(0.3)^2} = 53333333.33 \text{ Pa.s.m}^{-2}$

Again, Rate of filtration ($\frac{dV}{dt}$) = 5 × 10⁻⁵ m³/s

Time of filtration (t_n) = ?

Using equation 8.5 to calculate volume of filtrate (V_n) collected at time t_n, assuming 'L' to be negligible:

$$\frac{dV}{dt} = \frac{A^2 \Delta P}{\mu r w V_n}$$

Or, $V_n = \frac{A^2 \Delta P}{\mu r w \frac{dV}{dt}} = \frac{(0.2)^2 \times 200000}{53333333.33 \times 5 \times 10^{-5}} = 3 \text{ m}^3$

Hence, using equation 8.9, we can calculate t_n as:

$$t_n = \frac{\mu r w V^2}{2A^2 \Delta P} = \frac{53333333.33 \times (3)^2}{2 \times (0.2)^2 \times 200000} = 29999.99 s = 8.33 hr$$

Hence, the required time for filtration drop is 8.33 hr.

Exercises on rate of filtration

1. A slurry of concentration 23.5 Kg/m³ is filtered using a plate and frame filter press of 0.04 m² cross-sectional area. The filtrate viscosity is 8.9 × 10⁻⁴ Pa.s and the specific cake resistance is 1.9m/Kg. If the filtration is carried out at constant pressure drop of 340 KPa with a filter medium resistance of 10.6 × 10¹⁰ m⁻¹, calculate the time required to collect 60× 10⁻³ m³ of the filtrate.

2. A filtration is carried out for 10 min at a constant rate in a leaf filter and thereafter it is continued at constant pressure. This pressure is that attained at the end of the constant rate period. If one quarter of the total volume of the filtrate is collected during the constant rate period, what is the

total filtration time? Assume that the cake is incompressible and the filter medium resistance is negligible. [**Ans:** 85 minutes]

3. A test was carried out on a wine filter. It was found that under a constant pressure difference of 350 kPa gauge, the rate of flow was 450 kg h^{-1} from a total filter area of 0.82 m^2. Assuming that the quantity of cake is insignificant in changing the resistance of the filter, if another filter of 6.5 m^2 area is added, what pressure would be required for a throughput of 500 hectolitres per 8-hour shift from the combined plant? Firstly determine R the resistance, and then the pressure difference. Assume density of wine is 1000 m^{-3}.

Filter-cake compressibility

With some filter cakes, the specific resistance varies with the pressure drop across it. This is because the cake becomes denser under the higher pressure and so provides fewer and smaller passages for flow. The effect is spoken of as the compressibility of the cake. Soft and flocculent materials provide highly compressible filter cakes, whereas hard granular materials, such as sugar and salt crystals, are little affected by pressure. To allow for cake compressibility the empirical relationship has been proposed:

$$r = r'\Delta P^s \quad \ldots\ldots\ldots\ (8.10)$$

where r is the specific resistance of the cake under pressure P, ΔP is the pressure drop across the filter, r' is the specific resistance of the cake under a pressure drop of 1 atm and s is a constant for the material, called its compressibility.

This expression for r can be inserted into the filtration equations, such as equation. (8.6), and values for r' and *s* can be determined by carrying out experimental runs under various pressures.

Filtration equipments

The basic requirements for filtration equipment are:
- mechanical support for the filter medium,
- flow accesses to and from the filter medium and
- provision for removing excess filter cake.

In some instances, washing of the filter cake to remove traces of the solution may be necessary. Pressure can be provided on the upstream side of the filter, or a vacuum can be drawn downstream, or both can be used to drive the wash fluid through.

Plate and frame filter press

In the plate and frame filter press as shown in Fig. 8.3, a cloth or mesh is spread out over plates which support the cloth along ridges but at the same time leave a free area, as large as possible, below the cloth for flow of the filtrate. The plates with their filter cloths may be horizontal, but they are more usually hung vertically with a number of plates operated in parallel to give sufficient area.

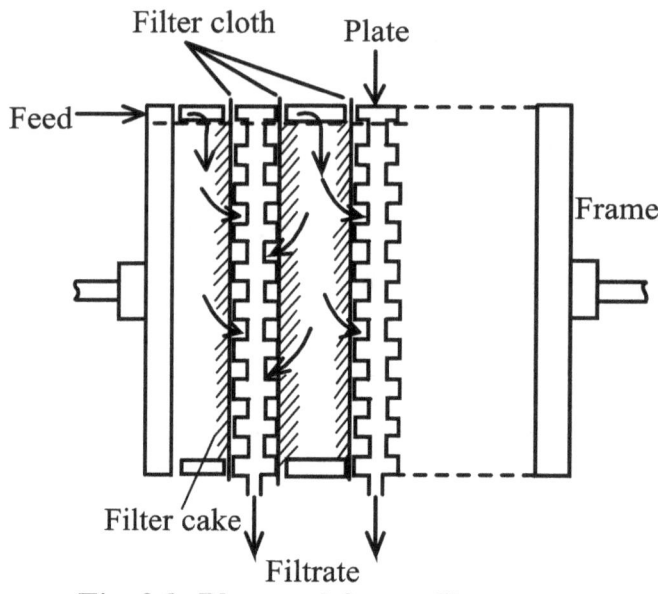

Fig. 8.3: Plate and frame filter press

Filter cake builds up on the upstream side of the cloth that is the side away from the plate. In the early stages of the filtration cycle, the pressure drop across the cloth is small and filtration proceeds at more or less a constant rate. As the cake increases, the process becomes more and more a constant-pressure one and this is the case throughout most of the cycle. When the available space between successive frames is filled with cake, the press has to be dismantled and the cake scraped off and cleaned, after which a further cycle can be initiated.

The plate and frame filter press is cheap but it is difficult to mechanize to any great extent. Variants of the plate and frame press have been developed which allow easier discharging of the filter cake. For example, the plates, which may be rectangular or circular, are supported on a central hollow shaft for the filtrate and the whole assembly enclosed in a pressure tank containing the slurry. Filtration can be done under pressure or vacuum. The advantage of vacuum filtration is that the pressure drop can be maintained whilst the cake is still under atmospheric pressure and so can be removed easily. The disadvantages are the greater costs of maintaining a given pressure drop by applying a vacuum and the limitation on the vacuum to about 80 kPa maximum. In pressure filtration, the pressure driving force is limited only by the economics of attaining the pressure and by the mechanical strength of the equipment.

Rotary filters

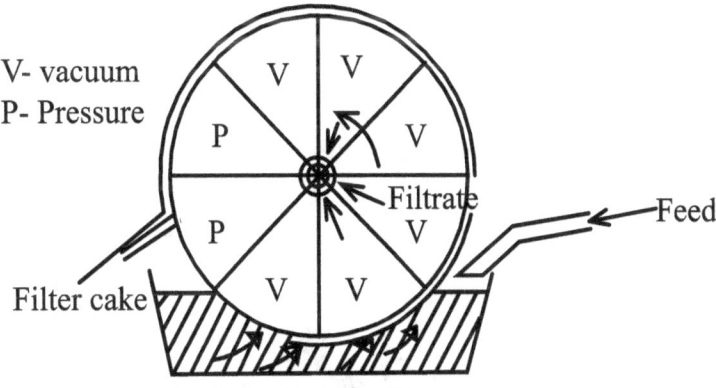

Fig. 8.4: Rotary filter

In rotary filters as given in Fig. 8.4, the flow passes through a rotating cylindrical cloth from which the filter cake can be continuously scraped. Either pressure or vacuum can provide the driving force, but a particularly useful form is the rotary vacuum filter. In this, the cloth is supported on the periphery of a horizontal cylindrical drum that dips into a bath of the slurry. Vacuum is drawn in those segments of the drum surface on which the cake is building up. A suitable bearing applies the vacuum at the stage where the actual filtration commences and breaks the vacuum at the stage where the cake is being scraped off after filtration. Filtrate is removed through trunnion bearings. Rotary vacuum filters are expensive, but they do provide considerable degree of mechanization and convenience.

Centrifugal filters

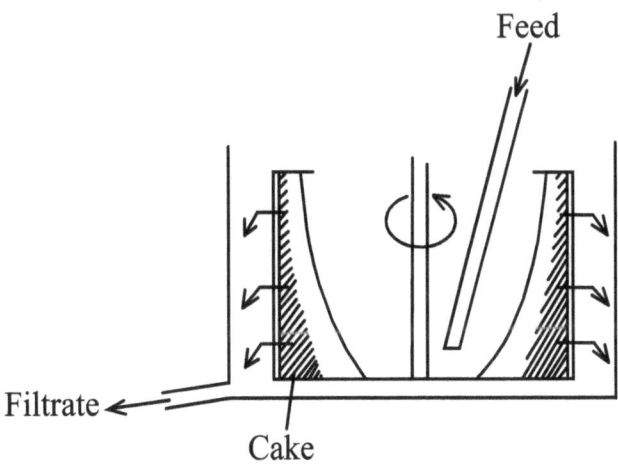

Fig. 8.5: Centrifugal filter

Centrifugal force is used to provide the driving force in some filters. These machines are really centrifuges fitted with a perforated bowl that may also have filter cloth on it. Liquid is fed into the interior of the bowl and under the centrifugal forces, it passes out through the filter material. If the object is rotated in a cylindrical cylinder, the contents of the fluid and solids exert an equal and opposite forces called the centrifugal force, outwards to the walls of the container. This is the force that causes settling or sedimentation of particles through a layer of liquid or filtration of a liquid through a bed of filter cake held inside a perforated rotating chamber.

Air filters

Fig. 8.6: Air filter

Filters are used quite extensively to remove suspended dust or particles from air streams. The air or gas moves through a fabric and the dust is left behind. These filters are particularly useful for the removal of fine particles. One type of bag filter consists of a number of vertical cylindrical cloth bags 15-30 cm in diameter, the air passing through the bags in parallel. Air bearing the dust enters the bags, usually at the bottom and the air passes out through the cloth. A familiar example of a bag filter for dust is to be found in the domestic vacuum cleaner. Some designs of bag filters provide for the mechanical removal of the accumulated dust. For removal of particles less than 5 mm diameter in modern air sterilization units, paper filters and packed tubular filters are used. These cover the range of sizes of bacterial cells and spores.

Chapter 9: Sedimentation

Introduction

Mechanical separations can be divided into four groups - sedimentation, centrifugal separation, filtration and sieving.

In sedimentation, two immiscible liquids, or a liquid and a solid, differing in density, are separated by allowing them to come to equilibrium under the action of gravity, the heavier material falling with respect to the lighter. This may be a slow process. It is often speeded up by applying centrifugal forces to increase the rate of sedimentation; this is called centrifugal separation. Filtration is the separation of solids from liquids, by causing the mixture to flow through fine pores which are small enough to stop the solid particles but large enough to allow the liquid to pass. Sieving, interposing a barrier through which the larger elements cannot pass, is often used for classification of solid particles.

The velocity of particles moving in a fluid

Under a constant force, for example the force of gravity, particles in a liquid accelerate for a time and thereafter move at a uniform velocity. This maximum velocity which they reach is called their terminal velocity. The terminal velocity depends upon the size, density and shape of the particles, and upon the properties of the fluid.

When a particle moves steadily through a fluid, there are two principal forces acting upon it, the external force causing the motion and the drag force resisting motion which arises from frictional action of the fluid. The net external force on the moving particle is applied force less the reaction force exerted on the particle by the surrounding fluid, which is also subject to the applied force, so that

$$F_s = V_a(\rho_p - \rho_f) \quad \ldots \ldots \ldots \text{ (9.1)}$$

where F_s is the net external accelerating force on the particle, V is the volume of the particle, a is the acceleration which results from the external force, ρ_p is the density of the particle and ρ_f is the density of the fluid.

The drag force on the particle (F_d) is obtained by multiplying the velocity pressure of the flowing fluid by the projected area of the particle

$$F_d = \frac{C\rho_f v^2 A}{2} \quad \ldots \ldots \ldots \text{ (9.2)}$$

where C is the coefficient known as the drag coefficient, ρ_f is the density of the fluid, v is the velocity of the particle and A the projected area of the particle at right angles to the direction of the motion.

If these forces are acting on a spherical particle so that $V = \pi D^3/6$ and $A = \pi D^2/4$, where D is the diameter of the particle, then equating F_s and F_d, in which case the velocity v becomes the terminal velocity v_m, we have:

$$\left(\frac{\pi D^3}{6}\right) \times a(\rho_p - \rho_f) = \frac{C\rho_f v_m^2 \pi D^2}{8} \quad \ldots\ldots\ldots (9.3)$$

It has been found, theoretically, that for the streamline motion of spheres, the coefficient of drag is given by the relationship:

$$C = \frac{24}{R_e} = \frac{24\mu}{D v_m \rho_f} \quad \ldots\ldots\ldots (9.4)$$

Substituting this value for C and rearranging, we arrive at the equation for the terminal velocity magnitude

$$\boxed{v_m = \frac{D^2 a(\rho_p - \rho_f)}{18\mu}} \quad \ldots\ldots\ldots (9.5)$$

This is the fundamental equation for movement of particles in fluids.

Sedimentation

Sedimentation uses gravitational forces to separate particulate material from fluid streams. The particles are usually solid, but they can be small liquid droplets, and the fluid can be either a liquid or a gas. Sedimentation is very often used in the food industry for separating dirt and debris from incoming raw material, crystals from their mother liquor and dust or product particles from air streams.

In sedimentation, particles are falling from rest under the force of gravity. Therefore in sedimentation, equation 9.5 takes the familiar form of Stokes' Law:

$$\boxed{v_m = \frac{D^2 g(\rho_p - \rho_f)}{18\mu}} \quad \ldots\ldots\ldots (9.6)$$

Note that equation 9.6 is not dimensionless and so consistent units must be employed throughout. For example, in the SI system D would be m, g in m s^{-2}, ρ in kg m^{-3} and μ in N s m^{-2}, and then v_m would be in ms^{-1}. Particle diameters are usually very small and are often measured in microns (micro-metres) = 10^{-6} m with the symbol μm.

Stoke's Law applies only in streamline flow and strictly only to spherical particles. In the case of spheres the criterion for streamline flow is that $(R_e) = 2$, and many practical cases occur in the region of streamline flow, or at least where streamline flow is a reasonable approximation.

Example 9.1: Settling velocity of dust particles

1. Calculate the settling velocity of dust particles of (a) 60 μm and (b) 10 μm diameter in air at 21°C and 100 kPa pressure. Assume that the particles are spherical and of density 1280 kg m^{-3}, and that the viscosity of air = 1.8×10^{-5} N s m^{-2} and density of air = 1.2 kg m^{-3}.

Sol^n:

Using equation 9.6;

a. For 60 μm particle:

$$v_m = \frac{(60 \times 10^{-6})^2 \times 9.81 \times (1280 - 1.2)}{(18 \times 1.8 \times 10^{-5})}$$

$$= 0.14 \text{ m s}^{-1}$$

b. For 10 μm particles since v_m is proportional to the squares of the diameters,

$$v_m = 0.14 \times \left(\frac{10}{60}\right)^2$$

$$= 3.9 \times 10^{-3} \text{ m s}^{-1}.$$

Using equation 2.5, checking the Reynolds number for the 60 μm particles,

$$R_e = \left(\frac{Dv\rho_b}{\mu}\right)$$

$$= \left(\frac{60 \times 10^{-6} \times 0.14 \times 1.2}{1.8 \times 10^{-5}}\right) = 0.56$$

Stokes' Law applies only to cases in which settling is free, that is where the motion of one particle is unaffected by the motion of other particles. Where particles are in concentrated suspensions, an appreciable upward motion of the fluid accompanies the motion of particles downward. So the particles interfere with the flow patterns round one another as they fall. Stokes' Law predicts velocities proportional to the square of the particle diameters. In concentrated suspensions, it is found that all particles appear to settle at a uniform velocity once a sufficiently high level of concentration has been reached. Where the size range of the particles is not much greater than 10:1, all the particles tend to settle at the same rate. This rate lies between the rates that would be expected from Stokes' Law for the largest and for the smallest particles. In practical cases, in which Stoke's Law or simple extensions of it cannot be applied, probably the only satisfactory method of obtaining settling rates is by experiment.

Free and hindered settling

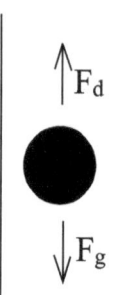

Fig. 9.1: Motion of particle in a fluid

For settling particles that are considered individually, i.e. dilute particle solutions, there are two main forces enacting upon any particle. The primary force is an applied force, such as gravity, and a drag force that is due to the motion of the particle through the fluid. The applied force is usually not affected by the particle's velocity, whereas the drag force is a function of the particle velocity.

For a particle at rest no drag force will be exhibited, which causes the particle to accelerate due to the applied force. When the particle accelerates, the drag force acts in the direction opposite to the particle's motion, retarding further acceleration, in the absence of other forces drag directly opposes the applied force. As the particle increases in velocity eventually the drag force and the applied force will approximately equate, causing no further change in the particle's velocity. This velocity is known as the terminal velocity, settling velocity or fall velocity of the particle. This is readily measurable by examining the rate of fall of individual particles.

The terminal velocity of the particle is affected by many parameters, i.e. anything that will alter the particle's drag. Hence the terminal velocity is most notably dependent upon grain size, the shape (roundness and sphericity) and density of the grains, as well as to the viscosity and density of the fluid.

Stokes, transitional and Newtonian settling describes the behavior of a single spherical particle in an infinite fluid, known as free settling. However this model has limitations in practical application. Alternate considerations, such as the interaction of particles in the fluid, or the interaction of the particles with the container walls can modify the settling behavior. Settling that has these forces in appreciable magnitude is known as hindered settling. Subsequently semi-analytic or empirical solutions may be used to perform meaningful hindered settling calculations.

Drag coefficient

In fluid dynamics, the drag coefficient (commonly denoted as: c_d, c_x or c_w) is a dimensionless quantity that is used to quantify the drag or resistance of an object in a fluid environment, such as air or water. It is used in the drag equation, where a lower drag coefficient indicates the object will have less aerodynamic or hydrodynamic drag. The drag coefficient is always associated with a particular surface area.

The drag coefficient of any object comprises the effects of the two basic contributors to fluid dynamic drag: skin friction and form drag. The drag coefficient of a lifting air foil or hydrofoil also includes the effects of lift-induced drag. The drag coefficient of a complete structure such as an aircraft also includes the effects of interference drag.

Sedimentation equipments
Simple gravity settling tank

Fig. 9.2: Settler for liquid- liquid dispersion

i. A simple gravity settler is shown in Fig. 9.2 for removing by settling a dispersed liquid phase from another phase. The velocity horizontally to the right must be slow enough to allow time for the smallest droplets to rise from the bottom to the interface or from the top down to the interference and coalescence.

ii. The dust- laden air enters at one end of a large, boxlike chamber called dust- settling chambers as shown in Fig. 9.3. Particles settle toward the floor at their terminal settling velocities. The air must remain in the chamber for a sufficient length of time so that the particles reach the floor of the chamber. Knowing the throughput of the air stream through the chamber and the chamber size, the residence time of the air in the chamber can be calculated.

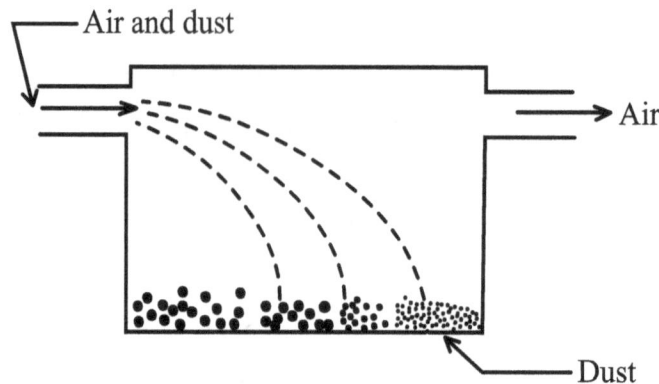

Fig. 9.3: Dust- settling chambers

The vertical height of the chamber must be small enough that this height, diameter by the settling velocities gives a time less than the residence time of the air.

Classifiers

i. Simple gravity settling classifier

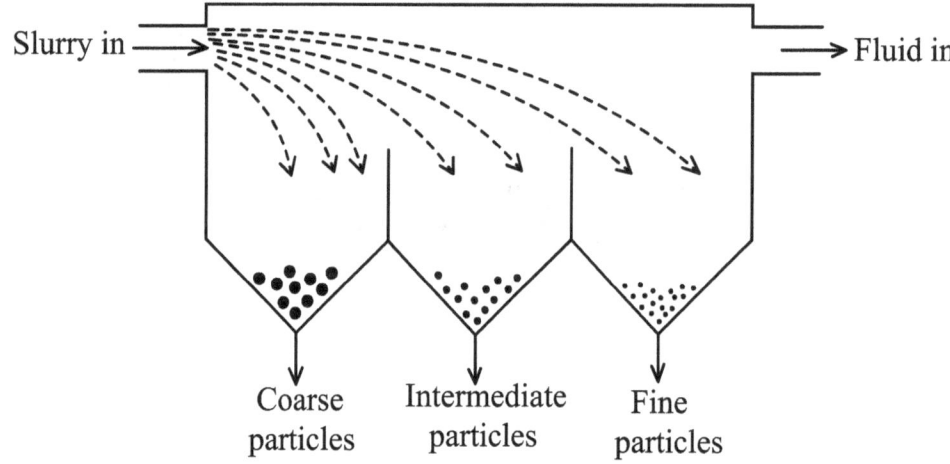

Fig. 9.4: Simple gravity settling classifier

The simplest type of classifier is one in which a large tank is sub-divided into several sections, as shown in the Fig. 9.4. A liquid slurry feed enters the tank containing a size range of solid particles. The larger, faster settling particles settle to the bottom close to the entrance and the slower- settling particles settle to the bottom close to the entrance and the slower settling particles settle to the bottom close to the exit. The linear velocity of the entering feed decreases as a result of the enlargement of the cross- sectional area at the entrance. The vertical baffles in the tank allow for the collection of several fractions.

156

ii. Spitzkasten classifier

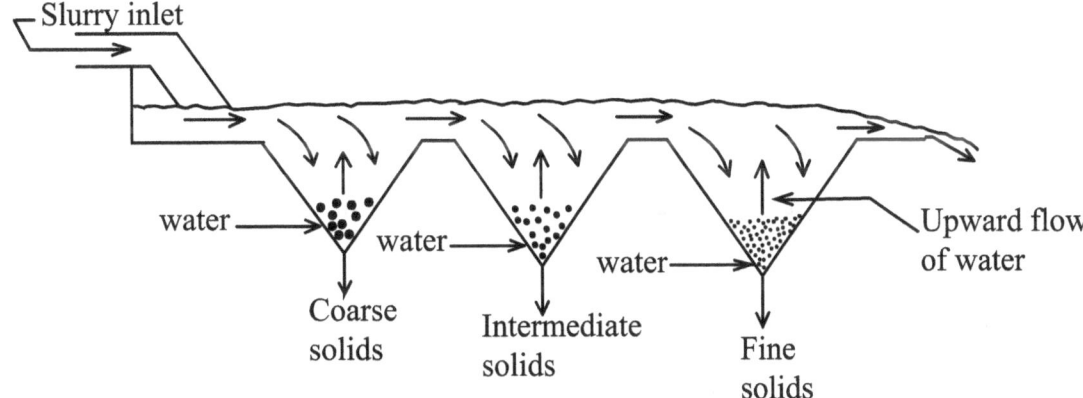

Fig. 9.5: Spitzkasten classifier

It consists of conical vessels of increasing diameter in the direction of flow. The slurry enters the first vessel where the largest and fastest- settling particles are separated. The overflow goes to the next vessel where another separation occurs. This continues in the succeeding vessel or vessels. In each vessel, the velocity of up flowing inlet water is controlled to give the desired size range for each vessel.

Sedimentation thickener

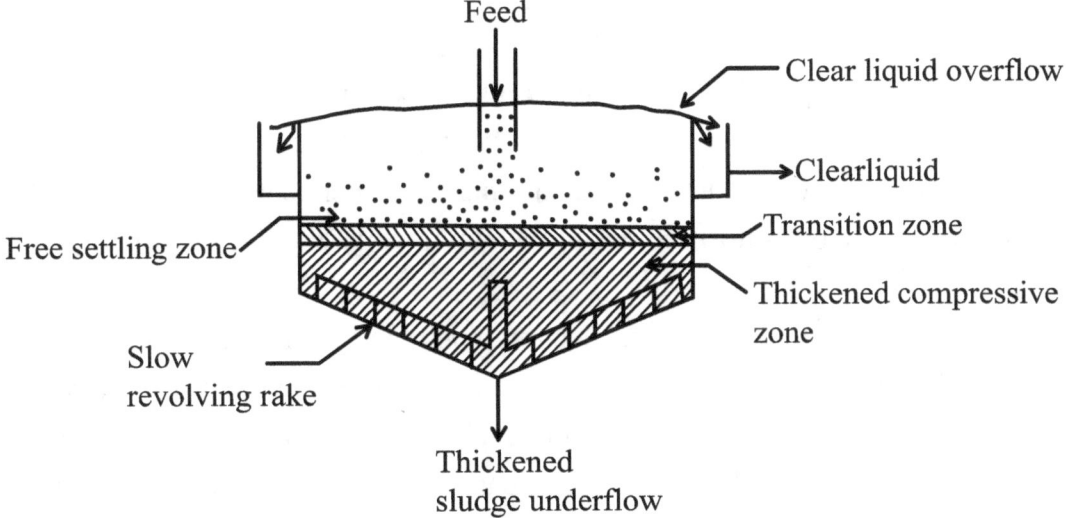

Fig. 9.6: Continuous thickener

The slurry is fed at the centre of the tank several feet below the surface of the liquid. Around the top edge of the tank is a clear liquid overflow outlet. The rake serves to scrap the sludge toward the centre of the bottom for removal. This gentle stirring aids in removing water from sludge.

In the thickener as shown in Fig. 9.6, the entering slurry spreads radially through the cross- section of the thickener and the liquid flows upward and the out the overflow. The solids settle in the upper zone by free settling. Below this dilute settling zone, there is the transition zone, in which the concentration of the solids increases rapidly and then the compression zone. A clear overflow can be obtained if the upward velocity of the fluid in the dilute zone is less than the minimal terminal settling velocity of the solids in this zone.

These settling rates are quite slow in the thickened zone, which consists of a compression of the solids with liquid being forced upward through the solids. This is an entrance case of hindered settling.

Centrifuge equipment

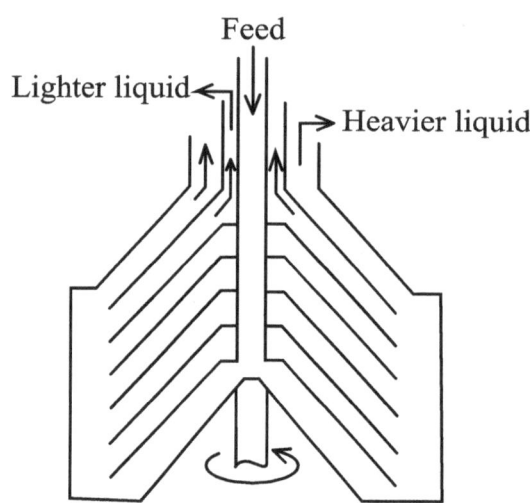

Fig. 9.7: Centrifugal cream separator

The simplest form of centrifuge consists of a bowl spinning about a vertical axis, as shown in Fig. 9.7. Liquids, or liquids and solids, are introduced into this and under centrifugal force the heavier liquid or particles pass to the outermost regions of the bowl, whilst the lighter components move towards the centre.

If the feed is all liquid, then suitable collection pipes can be arranged to allow separation of the heavier and the lighter components. Various arrangements are used to accomplish this collection effectively and with a minimum of disturbance to the flow pattern in the machine. To understand the function of these collection arrangements, it is very often helpful to think of the centrifuge action as analogous to gravity settling, with the various weirs and overflows acting in just the same way as in a settling tank even though the centrifugal forces are very much greater than gravity.

In liquid/liquid separation centrifuges, conical plates are arranged as illustrated in Fig. and these give smoother flow and better separation whereas liquid phases can easily be removed from a centrifuge, solids present much more of a problem.

Chapter 10: Size reduction

Introduction
Raw materials often occur in sizes that are too large to be used and, therefore, they must be reduced in size. This size-reduction operation can be divided into two major categories depending on whether the material is a solid or a liquid. If it is solid, the operations are called grinding and cutting, if it is liquid, emulsification or atomization. All depend on the reaction to shearing forces within solids and liquids.

Energies used in grinding
The energy "dE" required to produce a small change "dx" in the size of unit mass of material can be expressed as a power function of the size of material.

i.e. $$\boxed{\frac{dE}{dx} = -\frac{K}{x^n}} \quad \ldots\ldots\ldots (10.1)$$

Rittinger's law
Rittinger's considered that for the grinding of solids, the energy required should be proportional to the new surface produced and put n = 2.
Then,

$$\frac{dE}{dx} = -\frac{K}{x^2} \quad \ldots\ldots\ldots (10.2)$$

On integrating:

$$\int_0^E dE = -K.\int_{x_1}^{x_2} \frac{dx}{x^2}$$

$$\boxed{E = K\left(\frac{1}{x_2} - \frac{1}{x_1}\right)} \quad \ldots\ldots\ldots (10.3)$$

Where, x_1 is the average initial feed size, x_2 is the average final product size, E is the energy per unit mass required for the production of this new surface and is usually measured in Hp, and K is the Rittinger's constant which is found to hold better for fine grinding where a larger increase in surface results.

Kick's law

Kick considered that the energy required for a given size reduction was proportional to the size reduction ratio, which requires that n=1.

$$\text{Then,} \quad \frac{dE}{dx} = -\frac{K}{x} \quad \ldots\ldots\ldots (10.4)$$

On integrating:

$$\text{Or,} \quad \boxed{E = K \ln \frac{x_1}{x_2}} \quad \ldots\ldots\ldots (10.5)$$

where, x_1/x_2 being the size reduction ratio.

Kick's law has found to hold more accurate for coarser crushing where most of the energy is used in causing fracture along existing fissures. It gives the energy required to deform particles within the elastic limit. For many crushing operations, the energy requirement suggested by Kick's law appears to be too low, whereas that required by Rittinger's equation appears to be excessive.

Bond's law

It takes the value of n= 3/2 giving $\dfrac{dE}{dx} = -\dfrac{K}{x^{\frac{3}{2}}}$

On integrating:

$$\text{Or,} \quad \boxed{E = 2K \left(\frac{1}{\sqrt{x_2}} - \frac{1}{\sqrt{x_1}} \right)} \quad \ldots\ldots\ldots (10.6)$$

This third theory holds reasonably well for a variety of materials undergoing coarse, intermediate and fine grinding.

Work index

The work index (w_i) is defined as the gross energy required in kWh per ton of feed to reduce a very large feed to such a size that 80% of the product passes a 100mm screen.

If 80% of the feed passes a mesh size of D_{pb} millimetres and 80% of the product a mesh of D_{pa} millimetres, it follows that:

$$\frac{P}{m^*} = 0.3161 w_i \left(\frac{1}{\sqrt{D_{pb}}} - \frac{1}{\sqrt{D_{pa}}} \right) \quad \ldots\ldots\ldots (10.7)$$

Where P is the power required; m* is the feed rate to crusher; D_{pb} and D_{pa} are the screen apertures in mm.

Example 10.1: Milling of food using Rittinger's equation

1. Food is milled from 6 mm to 0.0012 mm using a 10 Hp motor. Would this motor be adequate to reduce the size of the particles to 0.0008 mm? Assume Rittinger's equation and that 1 Hp = 745.5 W.

Sol^n: Given,

Initial size of food (x_1) = 6mm = 6×10^{-3}m

Final size of food (x_2) = 0.0012mm = 0.0012×10^{-3}m

Energy of motor (E) = 10 Hp = 10×745.5W = 7455 W

Using Rittinger's equation, from 10.3;

$$E = K_R \left(\frac{1}{x_2} - \frac{1}{x_1} \right)$$

Or, $7455 = K_R \left\{ \left(\frac{1}{0.0012 \times 10^{-3}} \right) - \left(\frac{1}{6 \times 10^{-3}} \right) \right\}$; K_R= 0.0089

Therefore, Rittinger's constant (K_R) = 0.0089.

Again, to produce particle of 0.008 mm, using Rittinger's equation:

$$E = 0.0089 \left\{ \left(\frac{1}{0.0008 \times 10^{-3}} \right) - \left(\frac{1}{6 \times 10^{-3}} \right) \right\}$$

$$= 11.123 \text{ kW} = 15 \text{ Hp}$$

Hence, this motor is not adequate to reduce the size of the particles to 0.0008 mm.

Example 10.2: Crushing of limestone

2. What is the power required to crush 100 ton/h of limestone if 80% of the feed pass a 2-in screen and 80% of the product a 1/8 in screen? The work index for limestone is 12.74.

Sol^n: Given,

Feed rate to crusher (m*) =100 ton/h

Work index (w_i) =12.74

Size of screen aperture through which 80% feed passes (D_{pb}) =2 x 25.4=50.8 mm

Size of screen aperture through which 80% product passes (D_{pa}) =25.4/8=3.175 mm

Using equation 10.7, we can write:

$$\frac{P}{m^*} = 0.3161 w_i \left(\frac{1}{\sqrt{D_{pb}}} - \frac{1}{\sqrt{D_{pa}}} \right)$$

$$P = 100 \times 0.3162 \times 12.74 \times \left(\frac{1}{\sqrt{3.175}} - \frac{1}{\sqrt{50.8}} \right) = 169.6 \text{ KW}$$

Hence, the power required is 169.6 KW.

Exercises on energy required in size reduction

1. It is found that the energy required to reduce particles from a mean diameter of 1 cm to 0.3 cm is 11 kJ kg^{-1}. Estimate the energy requirement to reduce the same particles from a diameter of 0.1 cm to 0.01 cm assuming:
(a) Kick's Law, (b) Rittinger's Law, (c) Bond's Equation. [**Ans:** (a) 21 kJkg^{-1}; (b) 423 kJkg^{-1}; (c) 91 kJkg^{-1}]

2. A certain crusher accepts a feed of rock having diameter of 0.75 in and discharge a product of diameter 0.2 in. The power required to crush 12 tons per hr is 9.3 hp. What should be the power required if the capacity is reduced to 10 ton per hr and as a consequence of which the diameter of product become 0.15 in?

3. A continuous grinder obeying the Bond crushing law grinds a solid at the rate of 800 Kg/h from the initial diameter of 12 mm to a final diameter of 2 mm. If it is required to produce particles of 1 mm size, what would be the output rate of the grinder (in kg/h) for the same power input? [**Ans:** 470.32 Kg/h]

4. A material is crushed in a Blake jaw crusher such that the average size of particle is reduced from 50 mm to 10 mm, with the consumption of energy of 13.0 kW/ (kg/s). What will be the consumption of energy needed to crush the same material of average size 75 mm to average size of 25 mm: (a) assuming Rittinger's Law applies, (b) assuming Kick's Law applies? Which of these results would be regarded as being more reliable and why? [**Ans:** 4.33kW]

5. Sugar is ground from crystals of which it is acceptable that 80% pass a 500 mm sieve (US Standard Sieve No.35), down to a size in which it is acceptable that 80% passes a 88 mm (No.170) sieve, and a 5-horsepower motor is found just sufficient for the required throughput. If the

requirements are changed such that the grinding is only down to 80% through a 125 mm (No.120) sieve but the throughput is to be increased by 50% would the existing motor have sufficient power to operate the grinder? Assume Bond's equation. [**Ans:** 5.4 Hp]

Forces used in grinding

1. *Cutting*: The material is cut by means of sharp blades. Eg: Cutter mill
2. *Compression*: In this mode, the material is crushed between rollers by the application of pressure. Eg: Roller mill
3. *Impact*: This involves the operation of hammers or bars at high speeds. When a lump of material strikes the rotating hammers, the material splits apart. This action continues until particles of required sizes are obtained. Eg: Hammer mill

Impact also occurs when moving particles strike against a stationery surface. In the same way, particles moving at high speeds collide with each other and produce smaller particles. Eg: Fluid energy mill.

4. *Attrition*: This process involves breaking down of the material by rubbing action between the two surfaces i.e. Surface phenomena. Eg: Fluid energy mill.

Grinding equipments

Grinding equipment can be divided into two classes - crushers and grinders. In the first class the major action is compressive, whereas grinders combine shear and impact with compressive forces.

Crushers

Jaw and gyratory crushers are heavy equipment and are not used extensively in the food industry.

i. Jaw Crusher

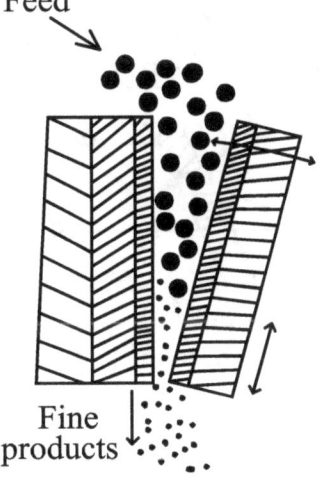

Fig. 10.1: Jaw crusher

In a jaw crusher, the material is fed in between two heavy jaws, one fixed and the other reciprocating, so as to work the material down into a narrower and narrower space, crushing it as it goes.

ii. Gyratory crusher

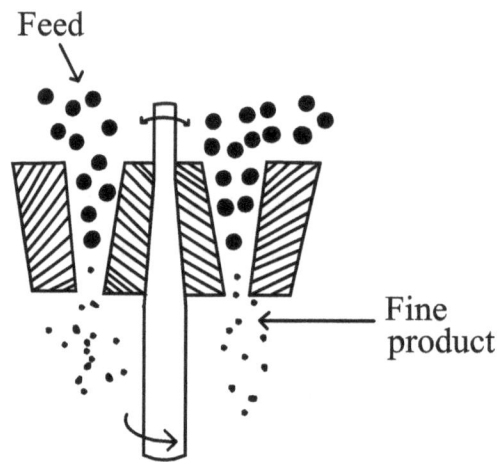

Fig. 10.2: Gyratory crusher

The gyrator crusher consists of a truncated conical casing, inside which a crushing head rotates eccentrically. The crushing head is shaped as an inverted cone and the material being crushed is trapped between the outer fixed, and the inner gyrating, cones, and it is again forced into a narrower and narrower space during which time it is crushed.

Grinders
i. Hammer mill

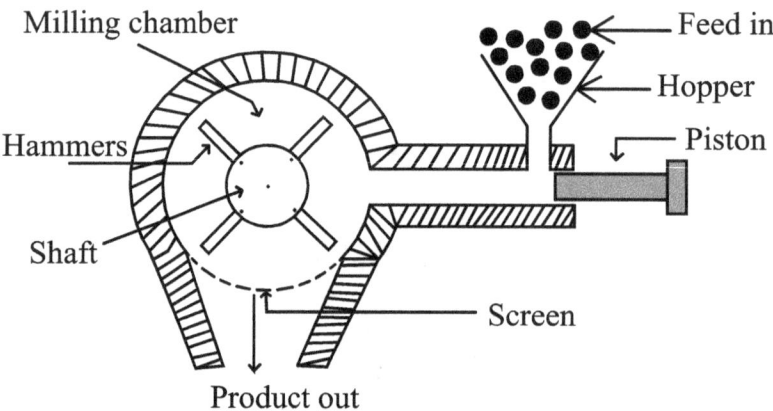

Fig. 10.3: Hammer mill

In a hammer mill, swinging hammerheads are attached to a rotor that rotates at high speed inside a hardened casing as shown in Fig. 10.3. It is used in the size reduction of sugar, tapioca, dry vegetables, extracted bones, dried milk, spices, pepper, etc

This type of impact or percussion grinder is common in the food industry. A high speed rotor carries a collar bearing a number of hammers around its periphery. When the rotor turns, the hammer heads swing through a circular path inside a closed fitting casing containing a toughened breaker plate. Feed passes into the action zone where the hammers drive the material against the breaker plate.

Reduction is mainly due to impact force, although attrition forces can also play a part in the size reduction. The hammers are often replaced by cutters, or by bars as in the beater bar mill. The hammer mill may be regarded as a general purpose mill, handling hard crystalline solids, fibrous materials, vegetable matter, sticky materials, etc. In the food industry, it is extensively used for grinding peppers and spices, dried milk, sugars, etc.

ii. Roller mill

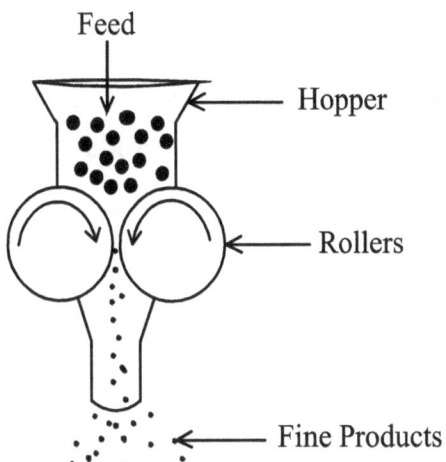

Fig. 10.4: Roller mill

Two parallel rolls rotate towards one another at high speeds, 400 – 3000rpm. Material is gravity fed between the parallel rolls and as it enters the 'nip' of the rollers, it experiences both crushing and shearing forces that grind the particle into smaller pieces. Roll surfaces can be either corrugated or smooth on the surface, depending on the need, and typically a speed differential between the rolls ranging from 1:1 to 3:1 is necessary. Due the low-impact grinding nature of roller mills, it is typically best suited for materials that break apart under pressure, otherwise known as 'friable' materials.

It is used in the size reduction of sugar cane, wheat, and chocolate refining.

iii. Disc attrition mill

These are of two types: a. Single disc mill .b. Double disc mill
It is used in the size reduction of Sugar, starch, cocoa powder, nutmeg, pepper, roasted nuts, cloves.

a. Single disc mill

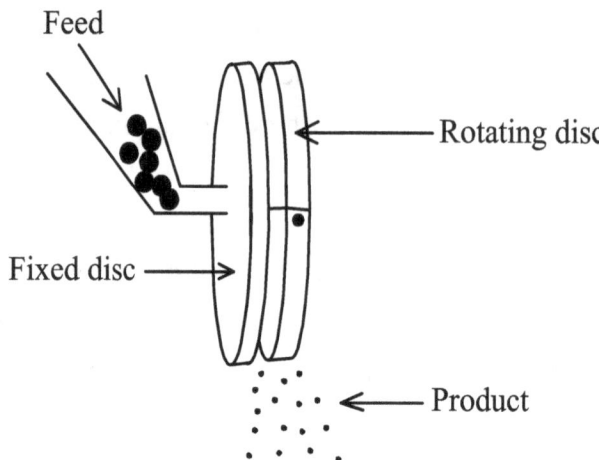

Fig. 10.5: Single disc attrition mill

In this device, the feed stock passes into a narrow gap between a high speed rotating grooved disc and the stationery casing of the mill. Intense shearing action results in comminution of the feed. The gap is adjustable, depending on the feed size and product requirements.

b. Double disc mill

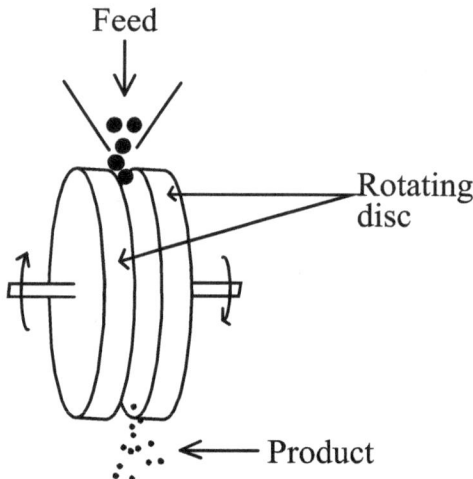

Fig. 10.6: Double disc attrition mill

In this modification, the casing contains two rotating discs. The disc rotate in opposite directions giving a greater degree of shear than that attainable in the single disc mill.

iv. Burr mill

This is an older type of disc attrition mill, originally used in flour milling. Two circular stones are mounted on a vertical axis. The upper stone, which is often fixed, has a feed entry port while the lower stone rotates.

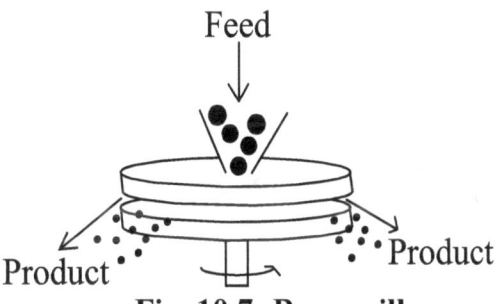

Fig. 10.7: Burr mill

The material, after subjection to the shearing force developed between the stones, is discharged over the edge of the lower stone. In some models, both the stones rotate, in opposite directions. This type of mill is being used in the wet milling of corn for the separation of starch gluten from the hulls.

v. Tumbling mill

It is of two types: a. Ball mill, and .b. Rod mill. It is used in the fining of food colors.

The only difference between the two is the use of high carbon stainless balls or rods in size reduction.

a. Ball mill

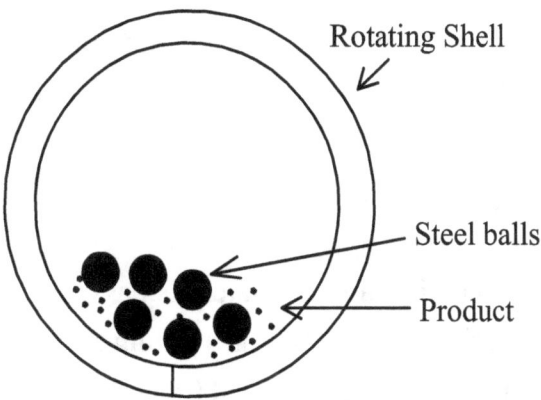

Fig. 10.8: Ball mill

In the ball mill as shown in Fig. 10.8, both the shearing and impact forces are utilized in the size reduction. The unit consists of horizontal, slow speed rotating cylinder containing a charge of steel balls or flint stones. As the cylinder rotates, balls are lifted up the sides of the cylinder and drop on to the material being comminuted which fills the void spaces between the balls. The balls also tumble over each other, exerting a shearing action on the feed material. This combination of impact and shearing forces brings about a very effective size reduction.

b. Rod mill

In the rod mill, the balls are replaced by high carbon steel rods. Impact and attrition forces still play a part but the effect of impact forces is less pronounced. Rod mills are recommended for use with sticky charges where balls can become trapped in the mass of the charge and become ineffective.

Screening

It is the unit operation in which a mixture of various sizes of solid particles is separated into two or more fractions by passing over a screen. Each fraction is more uniform in size than the original mixture.

A screen is a surface containing a number of equally sized apertures. The surface may be plane (horizontal or inclined) or it may be cylindrical. Small capacity plane screens are called sieves.

In general processing, Screening is widely used for separating mixtures of granulated and powdered materials into size ranges. When solid particles are dropped over a screen, the particles smaller than the size of screen openings pass through it, whereas larger particles are retained over the screen or sieve. A single screen can thus make separation into two fractions. When the feed is passed through a set of different sizes of sieves; it is separated into different fractions according to the size of openings of sieves.

Screening terminology

a. *Undersize, fines or minus* (-) materials are the materials passing through a given screen.

b. *Oversize, tails or Plus* (+) materials are the materials failing to pass a given screen. Either stream may be the desired (product) stream or the undersized (reject) stream, depending on the particular application.

c. *Screen aperture* is the space between the individual wires of a wire mesh screen. It is the minimum clear space between the edges of the opening in the screening surface and is generally specified in inches or millimetres.

d. *Mesh number* is the number of apertures per linear inch.

e. *Screen interval* is the relationship between successively decreasing opening in a standard screen series.

f. *The Fineness modulus (FM)* is an empirical figure obtained by adding the total % of the sample of an aggregate retained on each of a specified series of sieves, and dividing the sum by 100.

Fine aggregates range from a FM of 2.00 to 4.00, and coarse aggregates smaller than 38.1 mm range from 6.50 to 8.00. Combinations of fine and coarse aggregates have intermediate values.

g. *The Uniformity Index* is the ratio between large and small particles, multiplied by 100.

For example, a product whose UI is 50 means that the average small particle is half the size of the average large particle.

Different screen series

Tyler standard
This is widely used series based on a 200 mesh screen having 0.0021 inch diameter wires and a screen aperture of 0.0029 inch. The ratio between apertures in consecutive screen is 2 ½.

British standard
This screen series is based on wires following Standard wire gauge. A 200 mesh screen will have a screen aperture of 0.0030 inch and there is a screen interval of approximately 2 ¼ between the neighbouring screens.

U.S. bureau of standard
This series is based on 18 mesh screen with a 1 mm aperture and a screen interval of 2 ¼.

Exercises

1. Given the following sieve analysis:

Sieve size (mm)	% Retained
1.00	0
0.50	11
0.25	49
0.125	28
0.063	8
Through 0.063	4

plot a cumulative sieve analysis and estimate the weight fraction of particles of sizes between 0.300 and 0.350 mm and 0.350 and 0.400 mm. [**Ans:** 13% and 9%]

2. A sieve analysis gives the following results:
Plot a cumulative size analysis and a size-distribution analysis, and estimate the weights, per 1000 kg of powder, which would lie in the size ranges 0.150 to 0.200 mm and 0.250 to 0.350 mm.

Sieve size (mm)	Weight retained (g)
1.00	0
0.500	64
0.250	324
0.125	240
0.063	48
Through 0.063	24

Screen openings
Perforated metal screens

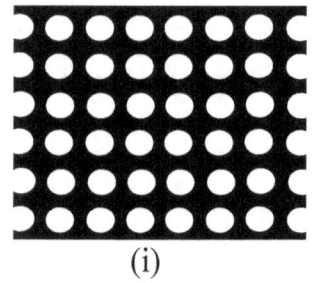 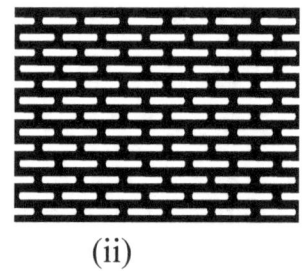 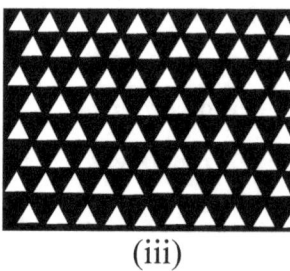

(i) (ii) (iii)

Fig. 10.9: Perforated metal screens with (i) Round openings (ii) Oblong openings (iii) Triangular openings

i. Round openings
The round opening in a perforated metal sheet are measured by the diameter (mm or inch) of the openings. Eg: 1/18 screen has round perforation of 1/18 inch in diameter or 2mm.

ii. Oblong opening
They are designed by two dimensions, the length and width of the opening. While mentioning oblong openings, the dimension of width is listed first and then the length as 1.8 ×20 mm.

iii. Triangular openings
There are two different systems used to measure triangular perforations. The most commonly used systems is to mention that the length of each side of the triangle in mm, it means 9mm triangle has 3 equal sides each 9mm long.

Wire mesh screens

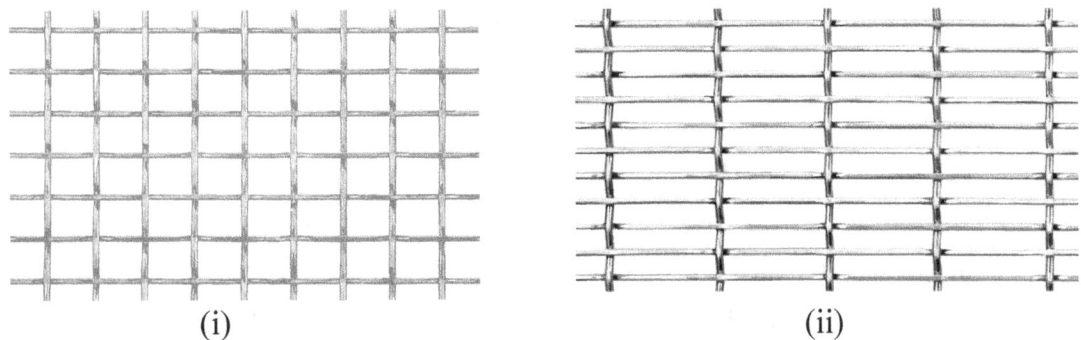

Fig. 10.10: Wire mesh screens with (i) Square openings (ii) Rectangular openings

i. Square mesh

The square opening in wire mesh are measured by the number of openings per inch in each direction. A 9 × 9 screen has 9 openings per inch.

ii. Rectangular mesh

The rectangular openings in wiremesh screens are measured in the same way as square wiremesh screen. A 3×6 rectangular wiremesh screen will have 3 openings per inch in one direction and 6 openings per inch in the other direction. The rectangles formed by the wire mesh are kept parallel to the direction of feed flow.

Factors affecting the efficiency of screening operation

Rate of feeding

If the rate of feeding is too high, insufficient residence time on the screening surface will result. The screen becomes overloaded and material capable of passing leaves with the oversize.

Particle size

Even though sufficiently small, the particle will only pass the screen if its alignment relative to the openings is favourable. Large particles tend to impede the passage of smaller and a preliminary separation may be required if a high proportion of bigger particles are present.

Moisture content

The presence of moisture in the feed can cause adhesion between small and large particles. Undersize particles will be removed with the oversized materials.

Worn or damaged screens

Oversized particles will pass through the damaged area and the efficiency of the separation will be impaired. Damaged screens should be repaired immediately. Fine screens are very fragile and should be treated with great care.

Blinding (clogging) of screens

Binding or clogging of the opening is particularly likely to occur when the size of the particles is near to that of the screen aperture. Particles capable of passing are prevented from doing so and are carried with the oversize. Blind screens should be cleaned immediately or screening efficiency will suffer.

Chapter 11: Mixing

Introduction

Mixing is the dispersing of components, one throughout the other. It occurs in innumerable instances in the food industry and is probably the most commonly encountered of all process operations. Unfortunately, it is also one of the least understood. There are, however, some aspects of mixing which can be measured and which can be of help in the planning and designing of mixing operations.

The mixing of liquids, solids and gases is one of the most common unit operations in the food industry. Mixing increases the homogeneity of a system by reducing non-uniformity or gradients in composition, properties or temperature. Secondary objectives of mixing include control of rates of heat and mass transfer, reactions and structural changes. In food processing applications, additional mixing challenges include sanitary design, complex rheology, desire for continuous processing and the effects of mixing on final product texture and sensory profiles. Mixing ensures delivery of a product with constant properties. For example, consumers expect all containers of soups, breakfast cereals, fruit mixes, etc to contain the same amount of each ingredient. If mixing fails to achieve the required product yield, quality, organoleptic or functional attributes, production costs may increase significantly.

Characteristics of mixtures

Ideally, a mixing process begins with the components, grouped together in some container, but still separate as pure components. Thus, if small samples are taken throughout the container, almost all samples will consist of one pure component. The frequency of occurrence of the components is proportional to the fractions of these components in the whole container. As mixing then proceeds, samples will increasingly contain more of the components, in proportions approximating to the overall proportions of the components in the whole container. Complete mixing could then be defined as that state in which all samples are found to contain the components in the same proportions as in the whole mixture.

Actually, this state of affairs would only be attained by some ordered grouping of the components and would be a most improbable result from any practical mixing process.

Another approach can then be made, defining the perfect mixture as one in which the components in samples occur in proportions whose statistical chance of occurrence is the same as that of a statistically random dispersion of the original components. Such dispersion represents the best that random mixing processes can do.

Measurement of mixing

The assessment of mixed small volumes, which can be taken or sampled, is what mixing measurement is all about. Sample compositions move from the initial state to the mixed state, and measurements of mixing must reflect this.

The problem at once arises, what size of sample should be chosen? To take extreme cases, if the sample is so large that it includes the whole mixture, then the sample composition is at once the mean composition and there remains no mixing to be done. At the other end of the scale, if it were possible to take samples of molecular size, then every sample would contain only one or other of the components in the pure state and no amount of mixing would make any difference. Between these lie all of the practical sample sizes, but the important point is that the results will depend upon sample size.

In many practical mixing applications, process conditions or product requirements prescribe suitable sample sizes. For example, if table salt is to contain 1% magnesium carbonate, the addition of 10 kg of magnesium carbonate to 990 kg of salt ensures, overall, that this requirement has been met.

However, if the salt is to be sold in 2 kg packets, the practical requirement might well be that each packet contains 20 g of magnesium carbonate with some specified tolerance, and adequate mixing would have to be provided to achieve this.

A realistic sample size to take from this mixture, containing 1000 kg of mixture, would be 2 kg. As mixing proceeds, greater numbers of samples containing both components appear and their composition tends towards 99% salt and 1% magnesium carbonate.

It can be seen from this discussion that the deviation of the sample compositions from the mean composition of the overall mixture represents a measure of the mixing process. This deviation decreases as mixing progresses. A satisfactory way of measuring the deviation is to use the statistical term called the standard deviation. This is the mean of the sum of the squares of the deviations from the mean, and so it gives equal value to negative and positive deviation and increasingly greater weight to larger deviations because of the squaring. It is given by:

$$s^2 = \frac{1}{n}\left\{(X_1 - \bar{X})^2 + (X_2 - \bar{X})^2 + \ldots + (X_n - \bar{X})^2\right\} \ldots\ldots\ldots (11.1)$$

where s is the standard deviation, n is the number of samples taken, $X_1, X_2, \ldots X_n$, are the fractional compositions of component X in the 1, 2, ... n samples and $\bar{x}$ is the mean fractional composition of component X in the whole mixture.

Using equation. 11.1, values of s can be calculated from the measured sample compositions, taking the n samples at some stage of the mixing operation. Often it is convenient to use s^2 rather than s, and s^2 is known as the variance of the fractional sample compositions from the mean composition.

Example 11.1: Mixing of binary mixtures

1. After a mixer mixing 99 kg of salt with 1 kg of magnesium carbonate had been working for some time, ten samples, each weighing 20 g, were taken and analysed for magnesium carbonate. The weights of magnesium carbonate in the samples were: 0.230, 0.172, 0.163, 0.173, 0.210, 0.182, 0.232, 0.220, 0.210, 0.213g. Calculate the standard deviation of the sample compositions from the mean composition.

Solution: Fractional compositions of samples, that is the fraction of magnesium carbonate in the sample, are respectively:

0.0115, 0.0086, 0.0082, 0.0087, 0.0105, 0.0091, 0.0116, 0.0110, 0.0105, 0.0107 (X)

Mean composition of samples, overall = 1/100 = 0.01 ($\bar{x}$)

Deviation of samples from mean, (0.0115 - 0.01), (0.0086 - 0.01), etc.

$$s^2 = \frac{1}{10}[(0.0115 - 0.01)^2 + (0.0086 - 0.01)^2 + \ldots] = 2.250 \times 10^{-6}$$

Hence, s = $\underline{1.5 \times 10^{-3}}$.

At some later time samples were found to be of composition: 0.0113, 0.0092, 0.0097, 0.0108, 0.0104, 0.0098, 0.0104, 0.0101, 0.0094, 0.0098, giving:

s = 3.7×10^{-7}.

and showing the reducing standard deviation. With continued mixing the standard deviation diminishes further.

The process of working out the differences can be laborious, and often the standard deviation can be obtained more quickly by making use of the mathematical relationship, proof of which will be found in any textbook on statistics:

$$s^2 = \frac{1}{n}\left\{\sum(X_1)^2 - \sum(\bar{x})^2\right\}$$

$$= \frac{1}{n}\left\{\sum(X_1)^2 - n(\bar{X})^2\right\}$$

$$= \frac{1}{n}\left\{\sum(X_1)^2\right\} - (\bar{X})^2 \quad \ldots\ldots\ldots (11.2)$$

If particles are to be mixed, starting out from segregated groups and ending up with the components randomly distributed, the expected variances (s^2) of the sample compositions from the mean sample composition can be calculated.

Consider a two-component mixture consisting of a fraction p of component P and a fraction q of component Q. In the unmixed state virtually all small samples taken will consist either of pure P or of pure Q. From the overall proportions, if a large number of samples are taken, it would be expected that a proportion p of the samples would contain pure component P. That is their deviation from the mean composition would be (1 - p), as the sample containing pure P has a fractional composition 1 of component P. Similarly, a proportion q of the samples would contain pure Q, that is, a fractional composition 0 in terms of component P and a deviation (0 - p) from the mean. Summing these in terms of fractional composition of component P and remembering that, p + q = 1.

$$s_o^2 = \frac{1}{n}\left\{pn(1-p)^2 + (1-p)n(0-p)^2\right\} \quad \text{(For n samples)}$$

$$= p(1-p)\ldots\ldots\ldots (11.3)$$

When the mixture has been thoroughly dispersed, it is assumed that the components are distributed through the volume in accordance with their overall proportions. The probability that any particle picked at random will be component Q will be q, and (1 - q) that it is not Q. Extending this to samples containing N particles, it can be shown, using probability theory, that:

$$s_r^2 = \frac{p(1-p)}{N} = \frac{s_o^2}{N} \quad \ldots\ldots\ldots (11.4)$$

This assumes that all the particles are equally sized and that each particle is either pure P or pure Q. For example, this might be the mixing of equal-sized particles of sugar and milk powder. The subscripts o and r have been used to denote the initial and the random values of s^2, and inspection of the formulae, equation 11.3 and equation 11.4 shows that in the mixing process the value of s^2 has decreased from p(1 - p) to 1/Nth of this value. It has been suggested that intermediate values between s_o^2 and s_r^2 could be used to show the progress of mixing. Suggestions have been made for a mixing index, based on this, for example:

$$(M) = \frac{(s_o^2 - s^2)}{(s_o^2 - s_r^2)} \quad \ldots\ldots\ldots (11.5)$$

which is so designed that (M) goes from 0 to 1 during the course of the mixing process. This measure can be used for mixtures of particles and also for the mixing of heavy pastes.

Example 11.2: Mixing index during dough mixing

For a particular bakery operation, it was desired to mix dough in 95 kg batches and then at a later time to blend in 5 kg of yeast. For product uniformity it is important that the yeast be well distributed and so an experiment was set up to follow the course of the mixing. It was desired to calculate the mixing index after 5 and 10 min mixing.

Sample yeast compositions, expressed as the percentage of yeast in 100 g samples were found to be:

After 5 min
   (%)   0.0 16.5 3.2 2.2 12.6 9.6 0.2 4.6 0.5 8.5
Fractional   0.0 0.165 0.032.......

After 10 min
   (%)   3.4 8.3 7.2 6.0 4.3 5.2 6.7 2.6 4.3 2.0
Fractional   0.034 0.083 ....

Solution:

Using the formula: $\frac{1}{n}\left\{\sum (X_i)^2\right\} - (\bar{X})^2$

And, calculating $s_5^2 = 3.0 \times 10^{-3}$
$s_{10}^2 = 3.8 \times 10^{-4}$

The value of $s_o^2 = 0.05 \times 0.95 = 4.8 \times 10^{-2}$

and $s_r^2 \approx 0$ as the number of "particles" in a sample is very large,

$$(M)_5 = \frac{(4.8 - 0.3)}{(4.8 - 0)} = \underline{0.93}$$

$$(M)_{10} = \frac{(4.8 - 0.04)}{(4.8 - 0)} = \underline{0.99}.$$

Hence, the mixing index after 5 mins and 10 mins of mixing are found to be 0.93 and 0.99 respectively.

Mixing of widely different quantities

The mixing of particles varying substantially in size or in density presents special problems, as there will be gravitational forces acting in the mixer which will tend to segregate the particles into size and density ranges. In such a case, initial mixing in a mixer may then be followed by a measure of (slow gravitational) un-mixing and so the time of mixing may be quite critical.

Mixing is simplest when the quantities that are to be mixed are roughly in the same proportions. In cases where very small quantities of one component have to be blended uniformly into much larger quantities of other components, the mixing is best split into stages, keeping the proportions not too far different in each stage. For example, if it were required to add a component such that its final proportions in relatively small fractions of the product are 50 parts per million, it would be almost hopeless to attempt to mix this in a single stage. A possible method might be to use four mixing stages, starting with the added component in the first of these at about 30:1. In planning the mixing process it would be wise to take analyses through each stage of mixing, but once mixing times had been established it should only be necessary to make check analyses on the final product.

Rates of mixing

Once a suitable measure of mixing has been found, it becomes possible to discuss rates of accomplishing mixing. It has been assumed that the mixing index ought to be such that the rate of mixing at any time, under constant working conditions such as in a well-designed mixer working at constant speed, ought to be proportional to the extent of mixing remaining to be done at that time. That is,

$$\frac{dM}{dt} = K[(1 - (M))] \quad \ldots\ldots\ldots \text{(11.6)}$$

where (M) is the mixing index and K is a constant, and on integrating from $t = 0$ to $t = t$ during which (M) goes from 0 to (M),

$$[(1 - (M))] = e^{-Kt}$$
$$\text{Or,} \boxed{(M) = 1 - e^{-Kt}} \quad \ldots\ldots\ldots \text{(11.7)}$$

This exponential relationship, using (M) as the mixing index, has been found to apply in many experimental investigations at least over two or three orders of magnitude of (M). In such cases, the constant K can be related to the mixing machine and to the conditions and it can be used to predict, for example, the times required to attain a given degree of mixing.

Example 11.3: Mixing time of binary mixtures

In a batch mixer, blending starch and dried-powdered vegetables for a soup mixture, the initial proportions of dried vegetable to starch were 40:60. If the variance of the sample compositions measured in terms of fractional compositions of starch was found to be 0.0823 after 300 s of mixing, for how much longer should the mixing continue to reach the specified maximum sample composition variance for a 24 particle sample of 0.02? Assume that the starch and the vegetable particles are of approximately the same physical size.

Sol^n:

Taking the fractional content of dried vegetables to be $p = 0.4$,

$$(1 - p) = (1 - 0.4) = 0.6$$
$$s_o^2 = 0.4 \times 0.6 = 0.24$$
$$s_r^2 = \frac{s_o^2}{N} = \frac{0.24}{24} = 0.01$$

Substituting in equation 11.5, we have:

$$(M) = \frac{(s_o^2 - s^2)}{(s_o^2 - s_r^2)}$$

$$= \frac{(0.24 - 0.0823)}{(0.24 - 0.01)} = 0.685$$

Substituting in equation 11.7;

$$e^{-300K} = 1 - 0.685 = 0.315$$
Or, $300K = 1.155$,
Hence, $K = 3.85 \times 10^{-3}$

For $s^2 = 0.02$,

$$(M) = \frac{(0.24 - 0.02)}{(0.24 - 0.01)} = 0.957$$

Therefore,

$$0.957 = 1 - e^{-0.00385t}$$
Or, $-0.043 = -e^{-0.00385t}$
Or, $3.147 = 0.00385t$
Hence, $t = 817s$, say 820 s,

The additional mixing time would be 820s - 300 s = 520 s.

Energy input in mixing

Quite substantial quantities of energy can be consumed in some types of mixing, such as in the mixing of plastic solids. There is no necessary connection between energy consumed and the progress of mixing: to take an extreme example there could be shearing along one plane in a sticky material, then recombining to restore the original arrangement, then repeating which would consume energy but accomplish no mixing at all. However, in well-designed mixers energy input does relate to mixing progress, though the actual relationship has normally to be determined experimentally. In the mixing of flour doughs using high-speed mixers, the energy consumed, or the power input at any particular time, can be used to determine the necessary mixing time. This is a combination of mixing with chemical reaction as flour components oxidize during mixing in air which leads to increasing resistance to shearing and so to increased power being required to operate the mixer.

Exercises on mixing index

1. Analysis of the fat content of samples from a chopped-meat mixture in which the overall fat content was 15% gave the following results, expressed as percentages:

    23.4 10.4 16.4 19.6 30.4 7.6

For this mixture, estimate the value of s^2, s_0^2, and s_r^2, if the samples are 5 g and the fat and meat are in 0.1 g particles in the mixture. [**Ans:** 0.68×10^{-2}; 12.8×10^{-2}; 0.256×10^{-2}]

2. If it is found that the mixture in Problem 1 is formed from the initial separate ingredients of fat and lean meat after mixing for 10 min, estimate the value of the mixing index after a further 5 min of mixing. [**Ans:** 0.966 ; 0.994]

Liquid mixing

Food liquid mixtures could in theory be sampled and analysed in the same way as solid mixtures but little investigational work has been published on this, or on the mixing performance of fluid mixers. Most of the information that is available concerns the power requirements for the most commonly used liquid mixer - some form of paddle or propeller stirrer. In these mixers, the fluids to be mixed are placed in containers and the stirrer is rotated. Measurements have been made in terms of dimensionless ratios involving all of the physical factors that influence power consumption. The results have been correlated in an equation of the form

$(P_o) = K(R_e)^n (F_r)^m$ .... **(11.8)**

where $(R_e) = (D^2N\rho/\mu)$, $(P_o) = (P/D^5N^3\rho)$ and this is called the Power number(relating drag forces to inertial forces), $(F_r) = (DN^2/g)$ and this is called the Froude number (relating inertial forces to those of gravity); D is the diameter of the propeller, N is the rotational frequency of the propeller (rev/sec), ρ is the density of the liquid, μ is the viscosity of the liquid and P is the power consumed by the propeller.

Notice that the Reynolds number in this instance uses the product DN for the velocity, which differs by a factor of p from the actual velocity at the tip of the propeller.

The Froude number correlates the effects of gravitational forces and it only becomes significant when the propeller disturbs the liquid surface. Below Reynolds numbers of about 300, the Froude number is found to have little or no effect, so that equation 11.8 becomes:

$$(P_o) = K(R_e)^n \quad \ldots\ldots\ldots (11.9)$$

Experimental results from the work of Rushton are shown plotted in Fig. 11.1.

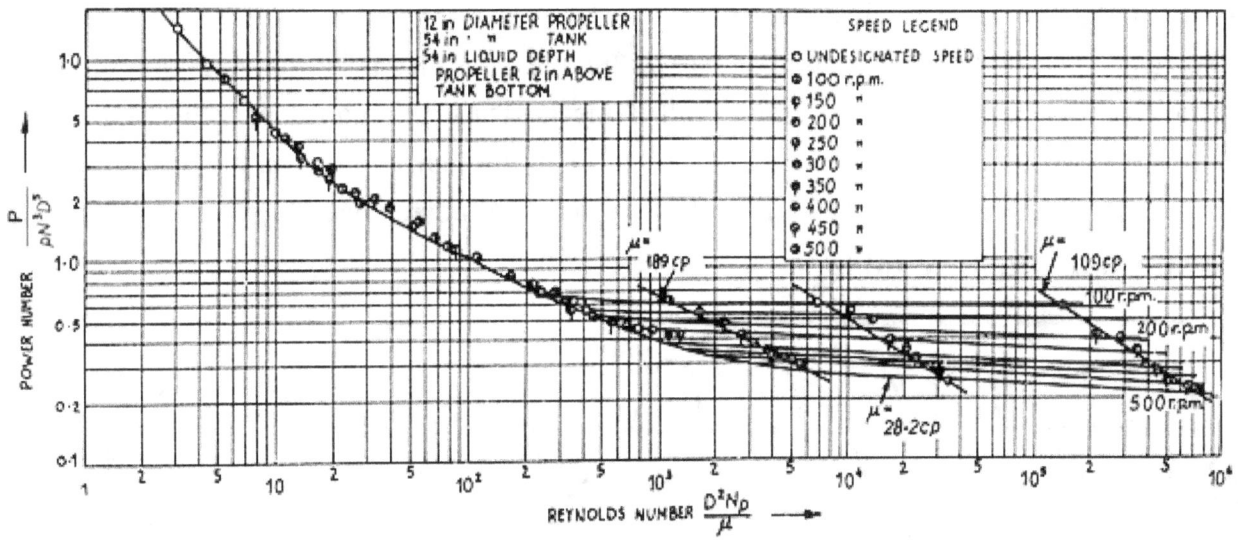

Fig. 11.1: Performance of propeller mixers

Unfortunately, general formulae have not been obtained, so that the results are confined to the particular experimental propeller configurations that were used. If experimental curves are available, then they can be used to give values for n and K in equation 11.8 and the equation then is used to predict power consumption. For example, for a propeller, with a pitch equal to the diameter, Rushton gives n = -1 and K = 41.

In cases in which experimental results are not already available, the best approach to the prediction of power consumption in propeller mixers is to use physical models, measure the factors, and then use equation 11.8 or equation 11.9 for scaling up the experimental results.

Types of mixing

There are 3 types of mixing:

i. Solid- solid mixing
ii. Liquid- liquid mixing
iii. Solid- liquid mixing.

Solid- solid mixing

In the mixing of particular solid material, the probability of getting an orderly arrangement of particles is virtually zero. The dynamic behavior of solid particles being mixed is complex.

The solid mixing is generally regarded as arising from one or more of the 3 basic mechanisms. They are: Convection, Diffusion and Shearing.

i. Convection: Transfer of masses or group of particles from one location to another.

ii. Diffusion: The transfer of individual particles from one location to another arising from the distribution of particles over a freshly developed surface.

iii. Shearing: The setting up of slipping planes within the mass also resulting in mixing of group of particles. Most mixing devices employ all 3 mechanisms although a particular type may predominate in any single mixer. Shear mixing is sometimes considered as a part of convection mechanism. Mixers: Tumbler mixer, Vertical screw mixer, Fluidized bed mixers.

a. Powder and particle mixers

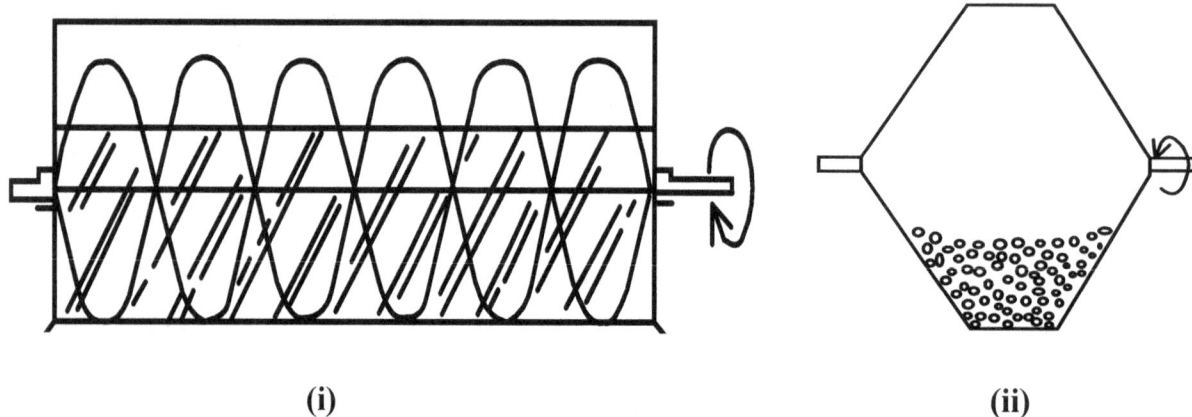

(i)   (ii)

Fig. 11.2: Mixers (i) ribbon blender (ii) double-cone mixer

The essential feature in these mixers is to displace parts of the mixture with respect to other parts. The ribbon blender, for example, shown in Fig. 2(i) consists of a trough in which rotates a shaft with two open helical screws attached to it, one screw being right-handed and the other left-handed. As the shaft rotates sections of the powder move in opposite directions and so particles are vigorously displaced relative to each other.

A commonly used blender for powders is the double-cone blender in which two cones are mounted with their open ends fastened together and they are rotated about an axis through their common base. This mixer is shown in Fig. 11.2(ii).

Solid- liquid mixing

One general principle which does not apply to the solid- liquid mixing (i.e. paste like material) is that their performances depend on direct contact between the mixing elements and the materials of the mixture. Thus the material must be brought to the mixing elements or the elements must travel to all the parts of the mixing vessel. The local action responsible for mixing have been described as kneading in which the materials pressed against other adjacent material or against the vessel walls and folding in which the fresh material is enveloped by already mixed materials. The material is subjected to shear and is often stretched and torn apart by the action of the mixing elements. In general, higher the consistency of the mixture, the greater the diameter of the impeller system and slower the speed of rotation.

The characteristics of solid- liquid phase which are to be mixed together are as follows:

i. Solid should not be coarse.

ii. Liquid should not be too viscous.

iii. The amount of solid per unit volume of liquid should not be too high.

a. Dough and paste mixers

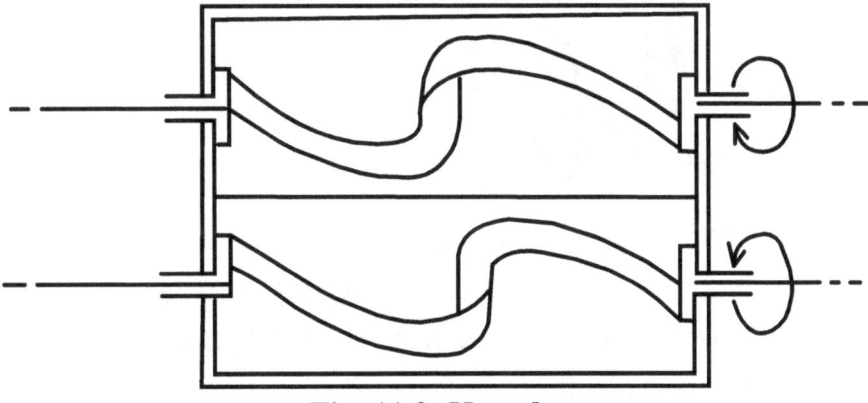

Fig. 11.3: Kneader

Dough and pastes are mixed in machines that have, of necessity, to be heavy and powerful. Because of the large power requirements, it is particularly desirable that these machines mix with reasonable efficiency, as the power is dissipated in the form of heat, which may cause substantial heating of the product. Such machines may require jacketing of the mixer to remove as much heat as possible with cooling water.

Perhaps the most commonly used mixer for these very heavy materials is the kneader which employs two contra-rotating arms of special shape, which fold and shear the material across a cusp, or division, in the bottom of the mixer. The arms are of so-called sigmoid shape as indicated in Fig. 11.3. They rotate at differential speeds, often in the ratio of nearly 3:2. Developments of this machine include types with multiple sigmoid blades along extended troughs, in which the blades are given a forward twist and the material makes its way continuously through the machine.

Another type of machine employs very heavy contra-rotating paddles, whilst a modern continuous mixer consists of an interrupted screw which oscillates with both rotary and reciprocating motion between pegs in an enclosing cylinder. The important principle in these machines is that the material has to be divided and folded and also displaced, so that fresh surfaces recombine as often as possible.

b. Homogenizers

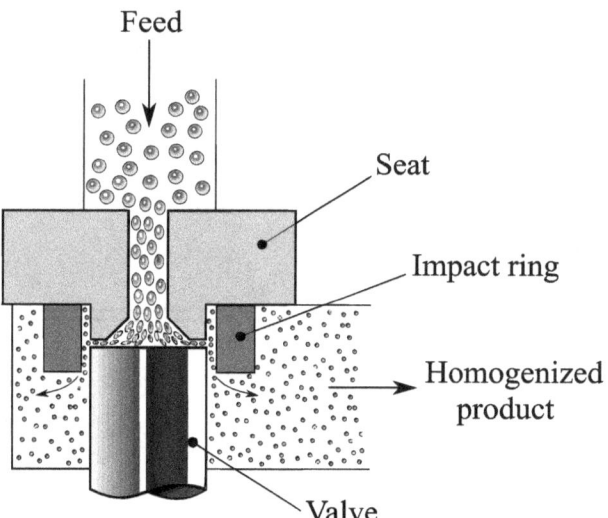

Fig. 11.4: Homogenizing valve

Homogenization refers to the process off forcing the milk through a homogenizer with the objectives off subdividing the fat globules. In effectively homogenized milk the fat globules are subdivided into microns or less in diameter.

Homogenized milk has been treated in such a manner so as to ensure break up of fat globules to such an extent that after 48 hours of quiescent storage at 45°F, no visible cream separation occurs in milk . The disrupting action takes place as a result of shearing action between the globules as they follow through a passage a high velocity. The solid particles nearest the edge of the stream are retarded somewhat by the friction of the fluid on the banks of the stream which therefore carries the particles nearest in the center at a more rapid velocity than those near the edge.

The difference in the speed causes the solid particles to grind against each other with shearing effect resulting in the reduction in size of the particles. The faster the flow and narrower the stream the greater is the shearing action.

Advantages of homogenization

1. No formation of cream layer.
2. Fat in the milk does not churn due to rough handling or excessive agitation.
3. More palatable, heavier body and richer flavor.
4. Produces soft curd and its better digested.
5. Less susceptible to oxidized flavor development.

Liquid- liquid mixing

Due to low diffusion capacity of liquid in comparison with air, external agency is required to get fast diffusion and hence mixing. So energy is supplied by means of agitation to obtain a homogenous mixture.

The mixture used for liquid – liquid mixing consists of one or more impeller, fixed in a rotating shaft which create currents within the liquid. These currents should travel throughout the mixing vessels. It is not sufficient simply to calculate the liquid. Turbulent condition must be created with the moving stream of liquid. When a moving stream of liquid comes into contact with the stationery or slowly moving liquid, shear occurs at the interface and low velocity liquid is entrained in the faster moving streams, mixing with the liquid there in. In order to achieve mixing in a reasonable time, the volumetric flowrate must be such that the entire volume of the mixing in a reasonable time.

The fluid velocity created by an impeller mixer in a tank has 3 components:

i. a radial velocity which acts in a direction perpendicular to the mixer shaft.

ii. a longitudinal velocity which acts parallel to the mixer shaft.

iii. a rotational velocity which acts tangential to the mixer shaft.

a. Liquid mixers

For the deliberate mixing of liquids, the propeller mixer as shown in Fig. 11.5 is probably the most common and the most satisfactory. In using propeller mixers, it is important to avoid regular flow patterns such as an even swirl round a cylindrical tank, which may accomplish very little mixing. To break up these streamline patterns, baffles are often fitted, or the propeller may be mounted asymmetrically.

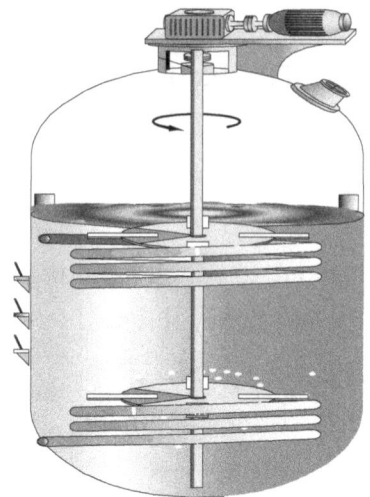

Fig. 11.5: Propeller mixer

Various baffles can be used and the placing of these can make very considerable differences to the mixing performances. It is tempting to relate the amount of power consumed by a mixer to the amount of mixing produced, but there is no necessary connection and very inefficient mixers can consume large amounts of power.

Factors influencing mixing

A single factor can in no way be considered as a unique indication of mixing. However flow properties of the components is the most important consideration which is again influenced by a number of factors:

Nature of the surface

Rough surface of one of the components does not induce satisfactory mixing. This can be due to the entry of active substance into the pores of the other ingredients.

Density of the particles

It is of minor importance. Demixing is accelerated when the density of the smaller particles is higher or when the mixing process is stopped abruptly. This is due to the fact that dense material always moves downward and settles at the bottom.

Particle size

It is easy to mix 2 powders having approximately the same particle size. The variation of the particle size can lead to separation because the small particles move downward through the spaces between the bigger particles. As the particle size increases, flow properties also increases due to the influence of the gravitational force on the size. Beyond a particular point, flow property decreases. The powder with a mean particle size of less than 100 μm are free flowing, which facilitates mixing.

Particle shape

The ideal particle is spherical in shape for the purpose of uniform mixing. The irregular shapes can become interlocked and there are less chances of separation of particles once these are mixed together.

Particle charge

Some particles exert attractive forces due to electrostatic charges on them. This can lead to separation or aggregation.

Proportion of materials

The best results can be obtained if two powders are mixed in equal proportions by weight and volume. If there is a large difference in the proportion of two powders, mixing is always done in the ascending order of their weights.

Index

A

Agitated batch crystallizers, 139
Attrition, 163

B

Ball mill, 167
Basket evaporator, 79
Batch process, 3
Bernoulli's equation, 28, 209
Bingham fluids, 32
Biot number, 57, 58
Boiling point diagram, 95, 97
boiling point elevation, 76
Bond's law, 160
British standard, 169
Burr mill, 166, 167

C

calandria, 79, 80, 85, 86
case hardening, 117
Climbing film long tube evaporator, 80
Compression, 163
constant rate period, 118, 120, 121, 146
Constant-pressure filtration, 144
Constant-rate filtration, 143
Continuous process, 3
critical moisture content, 118, 119, 121, 122
critical velocity, 19
crushers, 163

D

Dalton's law of partial pressure, 93
Darcy friction factor, 35
Differential distillation, 97, 98

dilatancy, 32
Disc attrition mill, 165
Discharge coefficient, 24
Double disc mill, 165, 166
drag coefficient, 152, 155
Drum dryers, 126
drying curve, 118, 119
Duhring's rule, 77
dust- settling chambers, 155

E

enriching section, 101, 102, 103, 104
Entrainment, 91
Equation of continuity, 9
equilibrium moisture content, 114, 119, 122

F

falling rate period, 119, 120
Fanning equation, 33, 34
Fanning friction factor, 34
film boiling, 66, 67
filter cake, 141, 142, 143, 144, 146, 147, 148, 149, 150
Flash or Equilibrium distillation, 97
Flow meters, 21
Flow work energy, 12
Fluidized bed dryers, 127, 128
Foaming, 76, 91
fouling coefficient, 65
Fourier number, 58
Fourier's law of heat conduction, 41
freeze drying, 75, 113
Froude number, 181

G

Grashof number, 59
grinders, 163
gyrator crusher, 164

H

hammer mill, 164, 165
heat exchanger, 68, 69, 70, 71, 72, 81
Henry's law, 94
hindered settling, 154, 158
Homogenizers, 184
Horizontal tube evaporator, 78

I

ideal solution, 93, 94, 96
Impact, 163, 168
Internal energy, 12

J

jaw crusher, 162, 164

K

Kick's law, 11, 160
Kinetic energy, 12
kneader, 184

L

Liquid- liquid mixing, 182, 185
Log mean temperature difference, 70
lumped heat capacitance, 57

M

magma, 129
mass flow rate, 9, 10, 11, 25
mixing index, 176, 177, 178, 180
Moody's diagram, 36
mother liquor, 129, 134, 139, 152

Multiple effect evaporator, 87

N

natural convection, 40, 58, 59, 60, 66, 68
Newtonian fluids, 32
Non-Newtonian fluids, 32
NTU method, 70
nucleate boiling, 66
Nusselt number, 59

O

Open pan evaporator, 78
orifice meter, 27
overall heat transfer coefficient, 62, 63, 64, 65, 70, 76, 83, 84, 85, 86

P

Package elevators, 193
Pitot tube, 28, 29, 30
plate and frame filter press, 146, 148
Plate evaporators, 81
Potential energy, 12
Power number, 181
Prandtl number, 59
pre-coat, 141, 142
pseudoplastic, 32

R

Raoult's law, 94, 96
reboiler, 101, 108, 109, 112
Rectification, 97, 100
reflux ratio, 110, 111, 112
Reynolds number, 18, 21, 34, 36, 153, 181
rheopectic, 32
Rittinger's law, 159
Rod mill, 167, 168
Roller mill, 163, 165

Rotameter, 30
rotary filters, 149

S

Screening, 168
Screw elevator, 193
Sedimentation thickener, 157
Semi-batch process, 4
settling down period, 118
sieving, 151
simple gravity settler, 155
Simple gravity settling classifier, 156
Single disc mill, 165, 166
Single effect evaporator, 87
Solid- liquid mixing, 182, 183
Solid- solid mixing, 182
Solubility, 75, 133, 134, 137
Spitzkasten classifier, 157
spray dryer, 126
Steady-state process, 3
steam economy, 87, 88, 90
Stokes' Law, 152, 153
streamline flow, 18, 36, 153
stripping section, 103, 104
surface heat transfer coefficient, 55, 61, 64

T

Tank crystallizers, 139
terminal velocity, 151, 152, 154

The Fineness modulus, 168
thermal conductivity, 41, 42, 43, 47, 48, 50, 51, 54, 55, 59, 61, 63, 64, 80
thixotropic, 32
transient flow, 19
transition boiling, 66
tray dryers, 123
Tumbling mill, 167
Tunnel dryers, 123
turbulent flow, 1, 17, 18, 19, 20
Tyler standard, 169

U

Uniformity Index, 169
unit operation, 1, 3, 4, 113, 168
Unsteady state heat transfer, 10, 39, 55
unsteady-state processes, 3

V

Vacuum crystallizers, 140
vacuum drying, 113
Venturimeter, 22, 25
Viscosity, 20, 31, 144, 146
volatility, 75, 94, 95, 107
volumetric flow rate, 9, 10, 11

W

work index, 160, 161

Appendices

Appendix 1: Unit and conversion factors

Length	1 inch	= 0.0254 m
	1 ft	= 0.3048 m
Area	1 ft^2	= 0.0929 m^2
Volume	1 ft^3	= 0.0283 m^3
	1 gal Imp	= 0.004546 m^3
	1 gal US	= 0.003785 m^3 = 3.785 litres
	1 litre	= 0.001 m^3
Mass	1 lb	= 0.4536 kg
	1 mole	molecular weight in kg
Density	1 lb/ft^3	= 16.03 kg m^{-3}
Velocity	1 ft/sec	= 0.3048 m s^{-1}
Pressure	1 lb/m^2	= 6894 Pa
	1 torr	= 133.3 Pa
	1 atm	= 1.013 x 10^5 Pa = 760 mm Hg
	1 Pa	= 1 N m^{-2} = 1 kg m^{-1} s^{-2}
Force	1 Newton	= 1 kg m s^{-2}
	1 lb ft s^{-2}	= 1.49 kg m s^{-2}
Viscosity	1 cP	= 0.001 N s m^{-2} = 0.001 Pa s
	1 lb/ft sec	= 1.49 N s m^{-2} = 1.49 kg m^{-1} s^{-2}
Energy	1 Btu	= 1055 J
	1 cal	= 4.186 J
Power	1 kW	= 1 kJ s^{-1}
	1 W	= 1 J s^{-1}
	1 horsepower	= 745.7 W = 745.7 J s^{-1} = 0.746 kW
	1 ton refrigeration	= 3.519 kW
Temperature units	(°F)	= 5/9 (°C) = 5/9 (K)
Heat-transfer coefficient	1 Btu ft^{-2} h^{-1} °F^{-1}	= 5.678 J m^{-2} s^{-1} °C
Thermal conductivity	1 Btu ft^{-1} h^{-1} °F^{-1}	= 1.731 J m^{-1} s^{-1} °C^{-1}
Constants	π	3.1416
	σ	5.73 x 10^{-8} J m^{-2} s^{-1} K^{-4}
	e (base of natural logs)	2.7183
	R	8.314 kJ mole^{-1} K^{-1} or 0.08206 m^3 atm mole^{-1} K^{-1}

(M) Mega = 10^6,
(k) kilo = 10^3,
(H) Hecto = 10^2
(m) milli = 10^{-3}
(μ) micro = 10^{-6}

Appendix 2: Some properties of solids (Atm. Pressure)

	Thermal conductivity ($J\ m^{-1}\ s^{-1}\ °C^{-1}$)	Specific heat ($kJ\ kg^{-1}\ °C^{-1}$)	Density ($kg\ m^{-3}$)	Temperature (°C)
1. Metals				
Aluminium	220	0.87	2640	0
Brass	97	0.38	8650	0
Cast iron	55	0.42	7210	0
Copper	388	0.38	8900	0
Steel, mild	45	0.47	7840	18
Steel, stainless	21	0.48	7950	20
2. Non-metals				
Asbestos sheet	0.17	0.84	890	51
Brick	0.7	0.92	1760	20
Cardboard	0.07	1.26	640	20
Concrete	0.87	1.05	2000	20
Celluloid	0.21	1.55	1400	30
Cotton wool	0.04	1.26	80	30
Cork	0.043	1.55	160	30
Expanded rubber	0.04		72	0
Fibreboard insulation	0.052		240	21
Glass, soda	0.52	0.84	2240	20
Ice	2.25	2.10	920	0
Mineral wool	0.04		145	30
Polyethylene	0.55	2.30	950	20
Polystyrene foam	0.036		24	0
Polyurethane foam	0.026		32	0
Polyvinyl chloride	0.29	1.30	1400	20
Wood shavings	0.09	2.5	1.50	0
Wood	0.28	2.5	700	30

Appendix 3: Some properties of gases (Atm. Pressure)

	Thermal conductivity $(J\ m^{-1}\ s^{-1}\ °C^{-1})$	Specific heat $(kJ\ kg^{-1}\ °C^{-1})$	Density $(kg\ m^{-3})$	Temperature $(°C)$
Ammonia	0.022	2.19	0.73	15
Carbon dioxide	0.015	0.80	1.98	0
Refrigerant 134a (tetrafluoroethane)		1.46	1.21	25
Ammonia	0.022	2.19	0.73	15
Nitrogen	0.024	1.005	1.3	0

Appendix 4: Some properties of liquids (Atm. Pressure)

	Thermal conductivity $(J\ m^{-1}\ s^{-1}\ °C^{-1})$	Specific heat $(kJ\ kg^{-1}\ °C^{-1})$	Density $(kg\ m^{-3})$	Viscosity $(N\ s\ m^{-2})$	Temperature $(°C)$
Water (see Appendix 6)					0
Sucrose 20% soln.	0.54	3.8	1070	1.92×10^{-3}	20
				0.59×10^{-3}	80
60% soln.				6.2×10^{-3}	20
				5.4×10^{-3}	80
				3.7×10^{-3}	20
Sodium chloride 22% soln.	0.54	3.4	1240	2.7×10^{-3}	20
Acetic acid	0.17	2.2	1050	1.2×10^{-3}	20
Ethyl alcohol	0.18	2.3	790	1.2×10^{-3}	20
Glycerine	0.28	2.4	1250	830×10^{-3}	20
Olive oil	0.17	2.0	910	84×10^{-3}	20
Rape-seed oil			900	118×10^{-3}	20
Soya-bean oil			910	40×10^{-3}	30
Tallow			900	18×10^{-3}	65
Milk (whole)	0.56	3.9	1030	2.2×10^{-3}	20
Milk (skim)			1040	1.4×10^{-3}	25
Cream 20% fat			1010	6.2×10^{-3}	3
30% fat			1000	$13,8 \times 10^{-3}$	3

Appendix 5: Thermal data for some food products

	Freezing point (°C)	Percent water	Specific heat (kJ kg^{-1}°C^{-1}) above freezing	Specific heat (kJ kg^{-1}°C^{-1}) below freezing	Latent heat of fusion (kJ kg^{-1})
Fruit					
Apples	-2	84	3.60	1.88	280
Bananas	-2	75	3.35	1.76	255
Grapefruit	-2	89	3.81	1.93	293
Peaches	-2	87	3.78	1.93	289
Pineapples	-2	85	3.68	1.88	285
Watermelons	-2	92	4.06	2.01	306
Vegetables					
Asparagus	-1	93	3.93	2.01	310
Beans (green)	-1	89	3.81	1.97	297
Cabbage	-1	92	3.93	1.97	306
Carrots	-1	88	3.60	1.88	293
Corn	-1	76	3.35	1.80	251
Peas	-1	74	3.31	1.76	247
Tomatoes	-1	95	3.98	2.01	310
Meat					
Bacon	-2	20	2.09	1.26	71
Beef	-2	75	3.22	1.67	255
Fish	-2	70	3.18	1.67	276
Lamb	-2	70	3.18	1.67	276
Pork	-2	60	2.85	1.59	197
Veal	-2	63	2.97	1.67	209
Miscellaneous					
Beer	-2	92	4.19	2.01	301
Bread	-2	32-37	2.93	1.42	109-121
Eggs	-3		3.2	1.67	276
Ice cream	-3 to -18	58-66	3.3	1.88	222
Milk	-1	87.5	3.9	2.05	289
Water	0	100	4.19	2.05	335

Appendix 6: Steam table - saturated steam

Temperature (°C)	Pressure(Absolute) (kPa)	Enthalpy (sat. vap.) (kJ kg^{-1})	Latent heat (kJ kg^{-1})	Specific volume (m^3 kg^{-1})
		Temperature Table		
0	0.611	2501	2501	206
1	0.66	2503	2499	193
2	0.71	2505	2497	180
4	0.81	2509	2492	157
6	0.93	2512	2487	138
8	1.07	2516	2483	121
10	1.23	2520	2478	106
12	1.40	2523	2473	93.9
14	1.60	2527	2468	82.8
16	1.82	2531	2464	73.3
18	2.06	2534	2459	65.0
20	2.34	2538	2454	57.8
22	2.65	2542	2449	51.4
24	2.99	2545	2445	45.9
26	3.36	2549	2440	40.0
28	3.78	2553	2435	36.6
30	4.25	2556	2431	32.9
40	7.38	2574	2407	19.5
50	12.3	2592	2383	12.0
60	19.9	2610	2359	7.67
70	31.2	2627	2334	5.04
80	47.4	2644	2309	3.41
90	70.1	2660	2283	2.36
100	101.4	2676	2257	1.67
105	120.8	2684	2244	1.42
110	143.3	2692	2230	1.21
115	169.1	2699	2217	1.04
120	198.5	2706	2203	0.892
125	232.1	2714	2189	0.771
130	270.1	2721	2174	0.669
135	313.0	2727	2160	0.582
140	361.3	2734	2145	0.509
150	475.8	2747	2114	0.393
160	617.8	2758	2083	0.307
180	1002	2778	2015	0.194
200	1554	2793	1941	0.127

Pressure Table

7.0	1.0	2514	2485	129
9.7	1.2	2519	2479	109
12.0	1.4	2523	2473	93.9
14.0	1.6	2527	2468	82.8
15.8	1.8	2531	2464	74.0
17.5	2.0	2534	2460	67.0
21.1	2.5	2540	2452	54.3
24.1	3.0	2546	2445	45.7
29.0	4.0	2554	2433	34.8
32.9	5.0	2562	2424	28.2
40.3	7.5	2575	2406	19.2
45.8	10.0	2585	2393	14.7
60.1	20.0	2610	2358	7.65
75.9	40.0	2637	2319	3.99
93.5	80.0	2666	2274	2.09
99.6	100	2676	2258	1.69
102.3	119	2680	2251	1.55
104.8	120	2684	2244	1.43
107.1	130	2687	2238	1.33
109.3	140	2690	2232	1.24
111.4	150	2694	2227	1.16
113.3	160	2696	2221	1.09
115.2	170	2699	2216	1.03
116.9	180	2702	2211	0.978
118.6	190	2704	2207	0.929
120.2	200	2707	2202	0.886
127.4	250	2717	2182	0.719
133.6	300	2725	2164	0.606
138.9	350	2732	2148	0.524
143.6	400	2739	2134	0.463
147.9	450	2744	2121	0.414
151.6	500	2749	2109	0.375
167.8	750	2766	2057	0.256
179.9	1000	2778	2015	0.194

Source: J. H. Keenan *et al.*, *Steam Tables - International Edition in Metric Units*, John Wiley, New York, 1969.

Appendix 7: Standard sieves

Aperture (m x 10^{-3})	ISO nominal aperture (m x 10^{-3})	U.S. no.	Tyler no.
22.6		7/8 in.	0.883 in.
16.0	16	5/8 in.	0.624 in.
11.2	11.2	7/16 in.	0.441 in.
8.0	8.00	5/16 in.	2 1/2 mesh
5.66	5.66	No.3 1/2	3 1/2 mesh
4.00	4.00	5	5 mesh
2.83	2.80	7	7 mesh
2.00	2.00	10	9 mesh
1.41	1.41	14	12 mesh
1.00	1.00	18	16 mesh
0.71	0.710	25	24 mesh
0.500	0.500	35	32 mesh
0.354	0.355	45	42 mesh
0.250	0.250	60	60 mesh
0.177	0.180	80	80 mesh
0.125	0.125	120	115 mesh
0.088	0.090	170	170 mesh
0.063	0.063	230	250 mesh
0.044	0.045	325	325 mesh

Note 500 μm = 0.50 m x 10^{-3} aperture = 35 US No. = 32 mesh

Appendix 8: Properties of saturated water

Temp. T, °C	Saturation Pressure P_{sat}, kPa	Density ρ, kg/m³ Liquid	Density ρ, kg/m³ Vapor	Enthalpy of Vaporization h_{fg}, kJ/kg	Specific Heat c_p, J/kg·K Liquid	Specific Heat c_p, J/kg·K Vapor	Thermal Conductivity k, W/m·K Liquid	Thermal Conductivity k, W/m·K Vapor	Dynamic Viscosity μ, kg/m·s Liquid	Dynamic Viscosity μ, kg/m·s Vapor	Prandtl Number Pr Liquid	Prandtl Number Pr Vapor	Volume Expansion Coefficient β, 1/K Liquid
0.01	0.6113	999.8	0.0048	2501	4217	1854	0.561	0.0171	1.792×10^{-3}	0.922×10^{-5}	13.5	1.00	-0.068×10^{-3}
5	0.8721	999.9	0.0068	2490	4205	1857	0.571	0.0173	1.519×10^{-3}	0.934×10^{-5}	11.2	1.00	0.015×10^{-3}
10	1.2276	999.7	0.0094	2478	4194	1862	0.580	0.0176	1.307×10^{-3}	0.946×10^{-5}	9.45	1.00	0.733×10^{-3}
15	1.7051	999.1	0.0128	2466	4185	1863	0.589	0.0179	1.138×10^{-3}	0.959×10^{-5}	8.09	1.00	0.138×10^{-3}
20	2.339	998.0	0.0173	2454	4182	1867	0.598	0.0182	1.002×10^{-3}	0.973×10^{-5}	7.01	1.00	0.195×10^{-3}
25	3.169	997.0	0.0231	2442	4180	1870	0.607	0.0186	0.891×10^{-3}	0.987×10^{-5}	6.14	1.00	0.247×10^{-3}
30	4.246	996.0	0.0304	2431	4178	1875	0.615	0.0189	0.798×10^{-3}	1.001×10^{-5}	5.42	1.00	0.294×10^{-3}
35	5.628	994.0	0.0397	2419	4178	1880	0.623	0.0192	0.720×10^{-3}	1.016×10^{-5}	4.83	1.00	0.337×10^{-3}
40	7.384	992.1	0.0512	2407	4179	1885	0.631	0.0196	0.653×10^{-3}	1.031×10^{-5}	4.32	1.00	0.377×10^{-3}
45	9.593	990.1	0.0655	2395	4180	1892	0.637	0.0200	0.596×10^{-3}	1.046×10^{-5}	3.91	1.00	0.415×10^{-3}
50	12.35	988.1	0.0831	2383	4181	1900	0.644	0.0204	0.547×10^{-3}	1.062×10^{-5}	3.55	1.00	0.451×10^{-3}
55	15.76	985.2	0.1045	2371	4183	1908	0.649	0.0208	0.504×10^{-3}	1.077×10^{-5}	3.25	1.00	0.484×10^{-3}
60	19.94	983.3	0.1304	2359	4185	1916	0.654	0.0212	0.467×10^{-3}	1.093×10^{-5}	2.99	1.00	0.517×10^{-3}
65	25.03	980.4	0.1614	2346	4187	1926	0.659	0.0216	0.433×10^{-3}	1.110×10^{-5}	2.75	1.00	0.548×10^{-3}
70	31.19	977.5	0.1983	2334	4190	1936	0.663	0.0221	0.404×10^{-3}	1.126×10^{-5}	2.55	1.00	0.578×10^{-3}
75	38.58	974.7	0.2421	2321	4193	1948	0.667	0.0225	0.378×10^{-3}	1.142×10^{-5}	2.38	1.00	0.607×10^{-3}
80	47.39	971.8	0.2935	2309	4197	1962	0.670	0.0230	0.355×10^{-3}	1.159×10^{-5}	2.22	1.00	0.653×10^{-3}
85	57.83	968.1	0.3536	2296	4201	1977	0.673	0.0235	0.333×10^{-3}	1.176×10^{-5}	2.08	1.00	0.670×10^{-3}
90	70.14	965.3	0.4235	2283	4206	1993	0.675	0.0240	0.315×10^{-3}	1.193×10^{-5}	1.96	1.00	0.702×10^{-3}
95	84.55	961.5	0.5045	2270	4212	2010	0.677	0.0246	0.297×10^{-3}	1.210×10^{-5}	1.85	1.00	0.716×10^{-3}
100	101.33	957.9	0.5978	2257	4217	2029	0.679	0.0251	0.282×10^{-3}	1.227×10^{-5}	1.75	1.00	0.750×10^{-3}
110	143.27	950.6	0.8263	2230	4229	2071	0.682	0.0262	0.255×10^{-3}	1.261×10^{-5}	1.58	1.00	0.798×10^{-3}
120	198.53	943.4	1.121	2203	4244	2120	0.683	0.0275	0.232×10^{-3}	1.296×10^{-5}	1.44	1.00	0.858×10^{-3}
130	270.1	934.6	1.496	2174	4263	2177	0.684	0.0288	0.213×10^{-3}	1.330×10^{-5}	1.33	1.01	0.913×10^{-3}
140	361.3	921.7	1.965	2145	4286	2244	0.683	0.0301	0.197×10^{-3}	1.365×10^{-5}	1.24	1.02	0.970×10^{-3}
150	475.8	916.6	2.546	2114	4311	2314	0.682	0.0316	0.183×10^{-3}	1.399×10^{-5}	1.16	1.02	1.025×10^{-3}
160	617.8	907.4	3.256	2083	4340	2420	0.680	0.0331	0.170×10^{-3}	1.434×10^{-5}	1.09	1.05	1.145×10^{-3}
170	791.7	897.7	4.119	2050	4370	2490	0.677	0.0347	0.160×10^{-3}	1.468×10^{-5}	1.03	1.05	1.178×10^{-3}
180	1,002.1	887.3	5.153	2015	4410	2590	0.673	0.0364	0.150×10^{-3}	1.502×10^{-5}	0.983	1.07	1.210×10^{-3}
190	1,254.4	876.4	6.388	1979	4460	2710	0.669	0.0382	0.142×10^{-3}	1.537×10^{-5}	0.947	1.09	1.280×10^{-3}
200	1,553.8	864.3	7.852	1941	4500	2840	0.663	0.0401	0.134×10^{-3}	1.571×10^{-5}	0.910	1.11	1.350×10^{-3}
220	2,318	840.3	11.60	1859	4610	3110	0.650	0.0442	0.122×10^{-3}	1.641×10^{-5}	0.865	1.15	1.520×10^{-3}
240	3,344	813.7	16.73	1767	4760	3520	0.632	0.0487	0.111×10^{-3}	1.712×10^{-5}	0.836	1.24	1.720×10^{-3}
260	4,688	783.7	23.69	1663	4970	4070	0.609	0.0540	0.102×10^{-3}	1.788×10^{-5}	0.832	1.35	2.000×10^{-3}
280	6,412	750.8	33.15	1544	5280	4835	0.581	0.0605	0.094×10^{-3}	1.870×10^{-5}	0.854	1.49	2.380×10^{-3}
300	8,581	713.8	46.15	1405	5750	5980	0.548	0.0695	0.086×10^{-3}	1.965×10^{-5}	0.902	1.69	2.950×10^{-3}
320	11,274	667.1	64.57	1239	6540	7900	0.509	0.0836	0.078×10^{-3}	2.084×10^{-5}	1.00	1.97	
340	14,586	610.5	92.62	1028	8240	11,870	0.469	0.110	0.070×10^{-3}	2.255×10^{-5}	1.23	2.43	
360	18,651	528.3	144.0	720	14,690	25,800	0.427	0.178	0.060×10^{-3}	2.571×10^{-5}	2.06	3.73	
374.14	22,090	317.0	317.0	0	—	—	—	—	0.043×10^{-3}	4.313×10^{-5}			

Source: Viscosity and thermal conductivity data are from J. V. Sengers and J. T. R. Watson, Journal of Physical and Chemical Reference Data 15 (1986), pp. 1291–1322. Other data are obtained from various sources or calculated.

Bibliography

1. AbdulHalim, R. G., Bhatt, P. M., Belmabkhout, Y., Shkurenko, A., Adil, K., Barbour, L. J., & Eddaoudi, M. (2017). A fine-tuned Metal-Organic Framework for Autonomous Indoor Moisture Control. Journal of the American Chemical Society. doi:10.1021/jacs.7b04132.
2. Andersen, S. A. (1959). Automatic Refrigeration. Nordborg: Danfoss.
3. Anonymous. (n.d.). Retrieved from http://www.thermaltransfersystems.com/pdf/alfa-laval-gasketed-heat-exchangers.Pdf.
4. Bansal, R. (2007). Engineering Mechanics: Laxmi Publications.
5. Birch, G. (1980). Principles of nutrition (4th edition). By Eva D. Wilson, Katherine H. Fisher and Pilar A. Garcia. John Wiley and Sons, New York, Chichester, Brisbane, Toronto, 1979. ix + 607 pp. Price: £9·50. Food Chemistry, 5(2), 188. doi:10.1016/0308-8146(80)90047-3.
6. Black, C., & Ditsler, D. E. (1974). Dehydration of Aqueous Ethanol Mixtures by Extractive Distillation. In D. P. Tassios (Ed.), Extractive and Azeotropic Distillation (Vol. 115, pp. 1–15). Washington, D. C.: American chemical society.
7. Blackadder, D. A. (1977). H. Leverne Williams: Polymer Engineering (Chemical Engineering Monographs Vol. 1). Chemical Engineering Science, 32(4), 459. doi:10.1016/0009-2509(77)85018-5.
8. BSc, P. C. B. (1971). Mechanical power transmission (1st ed.): The Macmillan Press Limited.
9. Chakraverty, A., & Singh, R. P. (2014). Postharvest technology and food process engineering: CRC Press.
10. Charm, S. E. (1971). The Fundamentals of Food Engineering (2nd ed. ed.). S.l.: Avi Publishing co.
11. Cooper, A. R., & Heuer, A. H. (Eds.). (1975). Mass Transport Phenomena in Ceramics. Boston, MA: Springer US.
12. DA, B., & RM, N. (1971). A Handbook of Unit Operations: London, England, Academic Press.
13. Daggett, W. L., Campbell, A. W., & Whitman, J. L. (1923). The electrometric titration of reducing sugars. Journal of the American Chemical Society, 45(4), 1043–1045. doi:10.1021/ja01657a025.

14. Dummies, C. (n.d.). Pressure, Speed and Bernoulli's equation in Physics Problems. Retrieved from http://www.dummies.com/how-to/content/pressure-speed-and-bernoullis-equation-in-physics-.html.
15. Earle, R. L. (2013). Unit operations in food processing: Elsevier.
16. Efficiency, B. o. E. (n.d.). Material and Energy Balance. Retrieved from http://www.emea.org/Guide%20Books/Book1/1.4%20MATERIAL%20%20AND%20ENERGY%20BALANCE.pdf.
17. Fellows, P. (Ed.) (2000). Food Processing Technology : Kirk-Othmer Encyclopedia of Chemical Technology. Hoboken, NJ, USA: CRC Press John Wiley & Sons, Inc.
18. Freeman, E. J. (1996). Streamflow Measurement. Environmental & Engineering Geoscience, II(4), 609–610. doi:10.2113/gseegeosci.II.4.609.
19. Geankoplis, C. J. (2003). Transport processes and separation process principles:(includes unit operations): Prentice Hall Professional Technical Reference.
20. Gutbier, A., & Weingärtner, E. (1913). Studien über Schutzkolloide. Erste Reihe: Stärke als Schutzkolloid. Kolloidchemische Beihefte, 5(6), 244–268. doi:10.1007/bf02558311.
21. H, R. W. (1950). An Introduction to Heat Transfer. M. Fishenden and O. A. Saunders. Oxford (Clarendon Press). 1950. 205 pp. Index. Diagrams. 15s. net. Journal of the Royal Aeronautical Society, 54(475), 486–487. doi:10.1017/s0368393100121522.
22. Hawkins, G. A. (1954). Heat Transmission. William H. McAdams. McGraw-Hill, New York-London, ed. 3, 1954. xiv + 532 pp. Illus. $8.50. Science, 120(3128), 984. doi:10.1126/science.120.3128.984.
23. Henderson, S. M., & Perry, R. L. (1976). Agricultural Process Engineering (3. ed. ed.). Westport, Conn.: AVI Publ.
24. Hunt, J. C. R. (1998). Qualitative Questions in Fluid Mechanics. In R. Moreau, A. Biesheuvel, & G. F. van Heijst (Eds.), Introduction to Fluid Mechanics : In Fascination of Fluid Dynamics (Vol. 45, pp. 483–501). Dordrecht: Elsevier Springer Netherlands.
25. Kern, D. Q. (1950). Process Heat Transfer (1. ed. ed.). New York [usw.]: McGraw-Hill.
26. Kofstad, P. (1975). Mass Transport Phenomena in Oxidation of Metals. In A. R. Cooper & A. H. Heuer (Eds.), Mass Transport Phenomena in Ceramics (pp. 383–407). Boston, MA: Springer US.
27. Lazar, T. (2003). Food Microbiology: Fundamentals and Frontiers (2nd edn). Trends in Food Science & Technology, 14(9), 390. doi:10.1016/s0924-2244(03)00047-5.

28. Learning, V. (n.d.). NPTEL problems. Retrieved from http://nptel.vtu.ac.in/VTU-NMEICT/HYD/problems.pdf.
29. Leatherhead Food International, L. F. I. (2008). Essential Guide to Food Additives. Cambridge: Royal Society of Chemistry.
30. Leniger, H. A., & Beverloo, W. A. (Eds.). (1975). Food Process Engineering: Dordrecht: Springer Netherlands.
31. Liu, D., & Liptak, B. (Eds.). (1997). Environmental Engineers' Handbook, Second Edition: CRC Press.
32. McMordie-Stoughton, K. L., Sandusky, W. F., Solana, A. E., & Bates, D. J. (2006). Updated Analysis of Energy and cost Savings for Utility service Program at Federal Sites.
33. Moreau, R., Biesheuvel, A., & van Heijst, G. F. (Eds.). (1998). Introduction to Fluid Mechanics : In Fascination of Fluid Dynamics. Dordrecht: Elsevier Springer Netherlands.
34. Olson, E. C. (1950). The temporal region of the Permian reptile Diadectes / Everett Claire Olson. [Chicago]: Chicago Natural History Museum.
35. Pursell, C. W. (1969). Christopher Colles's Steam Engine for the New York Water Works, 1775. Technology and Culture, 10(4), 567. doi:10.2307/3101577.
36. RB, K. (1978). Introduction to Industrial Drying Operations - With Helpful Charts for Following Drying Operations and a Large Number of Worked Examples. Oxford, England, Pergamon Press.
37. Reid, R. C. (1975). The McGraw-Hill dictionary of scientific and technical terms, Daniel N. Lapedes, Editor-in-Chief, McGraw-Hill Book Company, New York(1974).$39.50. AIChE Journal, 21(1), 206. doi:10.1002/aic.690210142.
38. Roberts, H. (1997). Air Pollution. In D. Liu & B. Liptak (Eds.), Environmental Engineers' Handbook, Second Edition: CRC Press.
39. Saito, H., Kubota, H., & Kunitomi, M. (1958). Fundamental Synthetic Conditions of Beryl by Hydrothermal Method. The Journal of the Society of Chemical Industry, Japan, 61(5), 545–548. doi:10.1246/nikkashi1898.61.545.
40. Sarbolouki, M. N. (1977). Reverse osmosis and synthetic membranes — theory, technology, engineering, S. Sourirajan, Ed., National Research Council of Canada, Ottawa, Canada, 1977, 598 pp. $45.00. Journal of Polymer Science: Polymer Letters Edition, 15(10), 629–630. doi:10.1002/pol.1977.130151011.

41. Shafer, W. H. (1990). Industrial Engineering and Operations Research. In W. H. Shafer (Ed.), Masters Theses in the Pure and Applied Sciences (pp. 273–287). Boston, MA: Springer US.
42. Simon, G. (1944). Book Review: When Thou Prayest . by E. G. Daunt. A.P.C.K., Dublin. 1s; Russian Letters of direction, 1834–1860. Trans. Iulia de Beausobre. Dacre Press. 5s When Thou Prayest . By DauntE. G.. A.P.C.K., Dublin. 1s; Russian Letters of direction, 1834–1860. Trans. de BeausobreIulia. Dacre Press. 5s. Theology, 47(293), 261–262. doi:10.1177/0040571x4404729308.
43. Singh, R. P., & Heldman, D. R. (2001). Introduction to Food Engineering (3. Aufl. ed.). s.l.: Elsevier professional.
44. Sodha, M. S. (1960). Transport Phenomena in Slightly Ionized Gases: High Electric Fields. Physical Review, 118(2), 378–381. doi:10.1103/PhysRev.118.378
45. Steinberg, S. H. (1957). The Statesman's Year-Book. Basingstoke: Palgrave Macmillan.
46. Strobeck, C. (1974). Outcrossing and Heterozygosity. Advances in Applied Probability, 6(1), 18. doi:10.2307/1426204
47. Swain, A. K., Patra, H., & Roy, G. K. (2011). Mechanical operations: Tata McGraw Hill.
48. Tassios, D. P. (Ed.) (1974). Extractive and Azeotropic Distillation. Washington, D. C.: American chemical society.
49. Toledo, R. T. (2007). Fundamentals of food process engineering: Springer Science & Business Media.
50. University, C. (2006). Collection of Solved Problems in Physics. Retrieved from https://physicstasks.eu/1162/the-venturi-tube
51. Van Arsdel, W. B. (1973). Food Dehydration (Ed. 2 ed.). Westport - Conn: Avi Publishing co.
52. Venkateshwaran, S. (1987). Mass Transfer and Reaction in Droplets. Ann Arbor: UMI.
53. Webb, R. L. (1992). A Review of: "Heat transfer design methods" John J. McKetta (ed.) Marcel-Dekker, New York, 1991, 612 pages, $165.00. Heat Transfer Engineering, 13(3), 83–85. doi:10.1080/01457639208939783
54. Zobel, M. (1976). Edward A. Kazarian: Food Service Facilities Planning. 230 Seiten, 64 Abb., 26 Tab., The AVI Publishing Co., Westport Connecticut 1975. preis: 17.00 $. Food / Nahrung, 20(6), 688. doi:10.1002/food.19760200633.

www.ingramcontent.com/pod-product-compliance
Lightning Source LLC
Chambersburg PA
CBHW080908170526
45158CB00008B/2043